AF692685

Ultra-Precision Machining of Difficult-to-Machine Materials

Ultra-Precision Machining of Difficult-to-Machine Materials

Guest Editor

Chen Li

MDPI

Basel • Beijing • Wuhan • Barcelona • Belgrade • Novi Sad • Cluj • Manchester

Guest Editor
Chen Li
School of Mechatronics
Engineering
Harbin Institute of Technology
Harbin
China

Editorial Office
MDPI AG
Grosspeteranlage 5
4052 Basel, Switzerland

This is a reprint of the Special Issue, published open access by the journal *Micromachines* (ISSN 2072-666X), freely accessible at: https://www.mdpi.com/journal/micromachines/special_issues/XR26NW5I5X.

For citation purposes, cite each article independently as indicated on the article page online and as indicated below:

Lastname, A.A.; Lastname, B.B. Article Title. *Journal Name* **Year**, *Volume Number*, Page Range.

ISBN 978-3-7258-6087-6 (Hbk)
ISBN 978-3-7258-6088-3 (PDF)
https://doi.org/10.3390/books978-3-7258-6088-3

© 2025 by the authors. Articles in this book are Open Access and distributed under the Creative Commons Attribution (CC BY) license. The book as a whole is distributed by MDPI under the terms and conditions of the Creative Commons Attribution-NonCommercial-NoDerivs (CC BY-NC-ND) license (https://creativecommons.org/licenses/by-nc-nd/4.0/).

Contents

About the Editor

Chen Li

Chen Li is a Full Professor at the School of Mechatronics Engineering, Harbin Institute of Technology. He completed his PhD studies in Aeronautics and Astronautics Manufacturing Engineering at the Harbin Institute of Technology. From 2018 to 2019, he worked as a visiting scholar at the University of Queensland, Australia. His interests include ultra-precision machining of advanced materials, composite energy field manufacturing, atomic-level manufacturing, nano-manufacturing, mechanical characterization, and the application of molecular dynamics and finite element and numerical simulations in machining technology. He has published more than 100 peer-reviewed research papers, and the total citations of these publications are more than 3500 in Google Scholar (h-index = 32). He serves as an Associate Editor of *International Journal of Extreme Manufacturing* and *Chinese Journal of Mechanical Engineering*.

Preface

Difficult-to-machine materials, such as semiconductors, engineering ceramics, optical glass, and composite materials, have been widely used in aerospace, integrated circuits, and energy power due to their superior mechanical properties and stable chemical characteristics. For advanced applications, these materials must be shaped into smooth substrates with high surface integrity using precision and ultra-precision machining technologies. However, these materials exhibit high brittleness, high hardness, anisotropic damage, and high elasticity, which present significant challenges for efficient machining. Severe surface and subsurface damage, along with serious cutting tool wear, are easily generated during the machining process, inevitably compromising the dimensional accuracy and service life of the components while increasing production costs. Revealing the damage evolution and material removal mechanism at micro- and nanoscales, exploring innovative machining technology, developing an innovative cutting tool, and optimizing machining process parameters are of great significance to realize the high efficiency and precision machining of difficult-to-machine materials. Traditional mechanical machining techniques primarily encompass grinding, lapping, polishing, turning, milling, drilling, etc. In recent years, with the development of high-energy beams and electrochemical processing technologies, composite energy field machining has increasingly demonstrated its advantages in machining difficult-to-machine materials in an efficient and ultra-precise manner. This Reprint focused on the precision machining of representative difficult-to-machine materials, including glassy carbon, high-strength steel, engineering ceramics, composite materials, monocrystalline silicon, and cemented carbide. Numerous innovative machining technologies were proposed in the Reprint, achieving significant results in material removal mechanisms, polishing, diamond wire saw cutting, ultrasonic vibration-assisted machining, laser-assisted machining, vibration-assisted electrode electrochemical drilling, and the fabrication of high-performance cutting tools. These findings effectively improved the surface quality of the workpiece and machining efficiency while reducing cutting forces and tool wear. These achievements laid a theoretical foundation and provided technical support for developing efficient and low-damage machining technologies for difficult-to-machine materials.

Chen Li
Guest Editor

micromachines

MDPI

Editorial

Editorial for Special Issue on Ultra-Precision Machining of Difficult-to-Machine Materials

Chen Li [1,2,3]

[1] School of Mechatronics Engineering, Harbin Institute of Technology, Harbin 150001, China; lichen1992@hit.edu.cn
[2] Suzhou Research Institute of HIT, Suzhou 215104, China
[3] Zhengzhou Research Institute, Harbin Institute of Technology, Zhengzhou 450046, China

1. Introduction for Special Issue of Ultra-Precision Machining of Difficult-to-Machine Materials

Difficult-to-machine materials, such as semiconductors [1,2], laser crystals [3,4], engineering ceramics [5,6], optical glass [7,8], superalloys [9,10], and composite materials [11–13], have been widely used in aerospace, integrated circuits, and energy power due to their superior mechanical properties and stable chemical characteristics. For instance, gallium nitride crystals are currently recognized as the most promising third-generation semiconductor material [14], with widespread applications in the aerospace and new energy sectors. YAG laser crystals serve as the primary host material for multi-kilowatt solid-state lasers [15]. Silicon carbide ceramics exhibit excellent specific stiffness and hardness, making them an ideal candidate material for fabricating large-aperture spaceborne reflectors [16]. Silicon carbide fiber-reinforced silicon carbide ceramic matrix composites are the preferred material for the thermal structural components of aircraft engines [17].

For advanced applications, these materials must be shaped into smooth substrates with high surface integrity using precision and ultra-precision machining technologies. Traditional mechanical machining techniques primarily encompass grinding [18], lapping [19], polishing [20,21], turning [22], milling [23], drilling [24], boring, etc. In recent years, with the development of high-energy beams and electrochemical processing technologies, composite energy field machining has increasingly demonstrated its advantages in machining difficult-to-machine materials in an efficient and ultra-precise manner. Examples of such technologies include laser-assisted machining [25,26], ultrasonic vibration-assisted machining [27], plasma-assisted machining [28], electrochemical–mechanical composite machining [29], and ion beam machining [30]. However, these materials exhibit high brittleness, high hardness, anisotropic damage [31], and high elasticity [32], which present significant challenges for efficient machining. Severe surface and subsurface damage, along with serious cutting tool wear, are easily generated during the machining process, inevitably compromising the dimensional accuracy and service life of the components while increasing production costs [33–35].

Understanding the mechanical properties [36,37], revealing the damage evolution and material removal mechanism at micro- and nanoscales [38,39], exploring innovative machining technology [40,41], developing an innovative cutting tool [42,43], and optimizing machining process parameters [44,45] are of great significance to realize the high efficiency and precision machining of difficult-to-machine materials. This Special Issue, "Ultra-Precision Machining of Difficult-to-Machine Materials", highlights recent research

 https://doi.org/10.3390/mi16091004

advancements in four key aspects within the machining field of difficult-to-machine materials, namely material removal mechanisms [46], abrasive machining technology [46–48], composite energy field machining technologies [49–53], and the development of high-performance cutting tools [54]. These advanced theories and technologies offer significant novel insights into efficient and low-damage machining of difficult-to-machine materials, comprising nine original articles.

2. Overview of Published Articles

2.1. Material Removal Mechanism Induced by Machining of Difficult-to-Machine Materials

Elucidating damage evolution and material removal mechanisms is essential for achieving the efficient and precise machining of difficult-to-machine materials. Jahnel et al. [46] investigated the brittle-to-ductile transition and the material removal mechanisms of glassy carbon induced by abrasive machining through nano-scratching and indentation tests. The results indicated that glassy carbon exhibited a distinct brittle-to-ductile transition behavior during the nanoscratching process, and the substrate underwent successive stages of ductile plastic deformation, funnel-shaped fracture, and brittle conchoidal fracture as the cutting depth increased. Appropriately reducing the feed rate and grinding depth while using abrasives with finer grain size is beneficial for improving the workpiece's surface quality. Inducing compressive stress during the machining process promotes more ductile material removal behavior due to the suppression of crack initiation and propagation. These findings not only enhance the understanding of material damage evolution and removal mechanisms at micro- and nanoscales but also provide a theoretical basis for achieving the high-efficiency and low-damage machining of brittle solids.

2.2. Abrasive Machining Technology of Difficult-to-Machine Materials

Abrasive machining technology involves the use of fine-grained abrasives or grinding wheels for workpiece precision machining. This technique represents a key method for achieving nanoscale ultra-smooth surface finishing of difficult-to-machine materials and primarily includes processes such as grinding, polishing, lapping, abrasive jet machining, and wire saw cutting. Flexible rheological polishing facilitates the fabrication of the smooth curved surfaces of GCr15 bushings. However, the inherent energy dissipation of the medium during flow leads to an uneven distribution of material removal. Zhao et al. [47] proposed controlling the physical field distribution of the abrasive medium through tool motion to achieve regulated material removal. In their study, the non-Newtonian behavior of the abrasive medium was characterized using a power law model. Based on principles of fluid dynamics and tribology, a film thickness model on the workpiece surface was developed. By solving this model, the pressure and velocity distributions within the film were determined. Based on contact mechanics, a single-abrasive-material removal model was developed, and a statistical method was employed to establish an abrasive distribution model, thereby enabling the formulation of a theoretical material removal model. Zhao refined this model through experimental validation, achieving a maximum error of no more than 14.0% and an average error of 11.1%. Guided by the corrected model, polishing a cylindrical inner surface for 60 min resulted in a surface roughness of 17.59 nm with a variance of 4.42 nm^2.

Zhang et al. [48] investigated the influence of diamond wire saw processing parameters on the surface characteristics of silicon nitride ceramics. Sawing experiments on silicon nitride ceramics were conducted within the range of processing parameters for diamond wire sawing, and the effects of cutting parameters on the surface morphology, surface roughness, and waviness of the as-sawn slices were analyzed. The results indicated that material removal on the diamond wire as-sawn surface of silicon nitride ceramics predominantly occurred in a brittle mode, as evidenced by the presence of brittle pits and regularly distributed wire marks within the 20–55 μm scale range. The surface roughness of the slices along the workpiece feed direction ranged from 0.27 to 0.38 μm and decreased with increasing wire speed and decreasing feed rate. Surface waviness varied between 0.09 and 0.21 μm, showing a consistent trend with the variation in sliced-surface roughness. The study results offer an experimental basis for advancing the engineering application of diamond wire sawing technology in machining silicon nitride ceramic components.

2.3. *Composite Energy Field Machining Technology of Difficult-to-Machine Materials*

Compared with traditional machining technologies, high-energy beam and electrochemical machining methods can effectively reduce machining forces, residual stresses, and tool wear, thereby demonstrating their advantages in the efficient and ultra-precision machining of difficult-to-machine materials. Numerous researchers have investigated the material removal mechanisms and process optimization strategies in composite energy field machining for such materials. Zhang et al. [49] proposed a longitudinal–torsional ultrasonic grinding (LTUG) process for machining the inner surface of GCr15 bushings. Modal analysis and amplitude testing were conducted to verify the structural rationality of the LTUG setup. Based on the probability density function of cutting thickness and the overlapping effect of adjacent abrasive trajectories, a surface topography prediction model for LTUG was established using the height formula of the surface residual material. The model's reliability was validated through orthogonal testing, achieving a prediction accuracy within 13.2%. According to the response surface methodology, the optimal process parameters were selected to satisfy the requirements of low surface roughness (Ra) and high material removal rate (MRR). The inner surface of the raceway plays a critical role in determining the performance of the bearing, and this study provides theoretical guidance for the longitudinal–torsional ultrasonic grinding process applied to the inner surface of GCr15 bearing steel.

Lin et al. [50] investigated the various fiber removal modes induced by orthogonal fiber weaving in SiCf/SiC composites. Force variations were analyzed through ultrasonic vibration-assisted scratching experiments to elucidate the influence of material removal mechanisms. In the study, three distinct surfaces, characterized by differences in fiber bundle weaving and lamination structures, were selected for comparative analysis. The results indicated that the internal fiber arrangement within the composite significantly affected the forces generated during the material removal process. Meanwhile, while investigating the influence of processing parameters on scratching forces, it was determined that the feed rate exerted the most significant effect. A comparative analysis between ultrasonic vibration-assisted scratching and traditional scratching experiments demonstrated that ultrasonic vibration can effectively reduce the scratching force. The underlying mechanism can be explained as follows: under the influence of ultrasonic assistance, the fibers experience brittle fracture, the matrix undergoes tearing, and surface residues are effectively removed in a timely manner. These combined effects contribute to a significant improvement in surface roughness after ultrasonic vibration-assisted processing.

Yang et al. [51] developed a novel anode vibration-assisted helical electrode electrochemical drilling method for fabricating micro-hole arrays on metal tube sidewalls in a highly efficient manner. Through comprehensive simulations and experimental investigations, the research team systematically examined the effects of the helical electrode rotation direction, rotational speed, and workpiece vibration parameters. The optimal results were obtained using forward electrode rotation at 3000 rpm in combination with workpiece vibration (8 μm amplitude at 100 Hz frequency), yielding precision micro-holes with a 200 μm diameter and a standard deviation of 3 μm. By employing this optimized process, the researchers successfully fabricated an array of 10 uniformly distributed micro-holes on 304 stainless steel tubes. Practical application tests demonstrated that the developed tube electrode enabled radial electrolyte flushing in electrochemical cutting, resulting in high-precision slits with an average width of 1.089 mm (standard deviation of 35.3 μm) on 5 mm thick stainless-steel plates. This breakthrough technology overcomes the critical limitations of conventional methods—such as drilling, laser cutting, or electrical discharge machining (EDM)—including thermal damage and burr formation, thereby offering an innovative solution for the efficient and precise machining of large-scale ruled surface components.

Cheng et al. [52] elucidated the mechanism underlying trajectory deviation and the uneven distribution of fracture quality in ceramic cutting using the thermal-controlled fracture method. Experimental results obtained from single-surface heating mode cutting demonstrate that the fracture trajectories on the upper and lower surfaces exhibit significant inconsistency, with the fracture quality being inferior to that achieved under dual-surface heating conditions. A finite element model was employed to calculate the stress distribution during the process, aiming to elucidate the underlying causes of the processing quality issues. This study demonstrates that trajectory deviation primarily results from the combined effects of lateral shear stress and transverse shear stress during ceramic cutting with a surface heat source. The uneven distribution of fracture quality in the single-surface heating mode is primarily attributed to the monotonic and highly gradient distribution of transverse tensile stress along the thickness direction of the workpiece. This study contributes to a deeper understanding of the processing challenges associated with this method, thereby facilitating the development of high-quality processing techniques in this field.

Wu et al. [53] conducted an investigation into quality control and damage mechanisms in the laser cutting of single-crystal silicon, systematically examining the regulatory effects of laser process parameters on cut seam dimensions and influenced processing zones in order to address the requirements of efficient wafer cutting. This study employed nanosecond pulsed laser cutting to process single-crystal silicon and innovatively applied surface erasure technology to eliminate the heat-affected zone. The results indicated that wiping the cutting surface effectively removed the processing-affected zones and recast layers on both sides of the cut seam, restoring the surface to a condition closely resembling its original state. Variations in the number of cutting passes influenced the surface morphology of the groove, with the optimal surface morphology observed at 20 cuts. This study quantitatively analyzed the correlation between laser parameters and damage morphology, demonstrating that optimized parameter settings can simultaneously achieve high cutting quality and minimal thermal damage, thereby offering a novel process approach for semiconductor manufacturing.

2.4. Development of High-Performance Cutting Tools

Cutting tools are often referred to as the "industrial teeth" of machine tools, and their performance directly influences production efficiency and machining quality. With the growing demand for high-precision and high-efficiency machining in sectors such as aerospace, increasingly stringent requirements have been placed on the performance of cutting tools. The development of technologies for fabricating high-performance cutting tools plays a crucial role in achieving the efficient and precise machining of difficult-to-cut materials. Qin et al. [54] proposed a chemical–mechanical synergistic preparation (CMSP) method for the cutting edge of cemented carbide inserts. A CMSP device was specifically designed and constructed to process insert cutting edges. Subsequently, the Taguchi method was integrated with gray relational analysis and fuzzy inference to optimize the polishing slurry formulation used in the CMSP process for insert cutting edges. In addition, orthogonal experiments, the Taguchi method, and analysis of variance (ANOVA) were employed to evaluate the effects of the polishing plate's rotational speed, swing angle, and controller input frequency on the edge preparation process and to optimize the corresponding parameters. The results indicated that the optimal parameter combination for the polishing slurry used in the processing of cemented carbide inserts included an abrasive particle mass concentration of 10 wt%, an oxidant mass concentration of 10 wt%, a dispersant mass concentration of 2 wt%, and a pH value of 8. For the linear edge CMSP process, the optimal parameter combination consisted of a polishing plate rotational speed of 90 rpm, a swing angle of 6°, and a controller input frequency of 5000 Hz. The optimal CMSP process parameter combination for the circular edge included a polishing plate rotational speed of 90 rpm, a swing angle of 6°, and a controller input frequency of 7000 Hz. Among these parameters, the polishing plate rotational speed exerted the most significant influence on the edge preparation process, followed by the swing angle, while the controller input frequency had the least impact. This study demonstrated that CMSP represents a promising approach for treating cemented carbide insert cutting edges in tool manufacturing enterprises.

3. Conclusions

The articles published in this Special Issue focus on the precision machining of representative difficult-to-machine materials, including glassy carbon [46], high-strength steel [47,49,51], engineering ceramics [48,52], composite materials [50], monocrystalline silicon [53], and cemented carbide [54]. Authors have proposed numerous innovative machining technologies and process optimizations for the aforementioned difficult-to-machine materials, achieving significant results in material removal mechanisms, polishing, diamond wire saw cutting, ultrasonic vibration-assisted machining, laser-assisted machining, vibration-assisted electrode electrochemical drilling, and the fabrication of high-performance cutting tools. These research findings effectively improved the surface quality of the workpiece and machining efficiency while reducing cutting forces and tool wear. These achievements laid a theoretical foundation and provided technical support for developing efficient and low-damage machining technologies for difficult-to-machine materials.

Author Contributions: Writing—original draft: C.L.; review: C.L. All authors have read and agreed to the published version of the manuscript.

Conflicts of Interest: The authors declare no conflicts of interest.

References

1. Li, C.; Piao, Y.; Meng, B.; Hu, Y.; Li, L.; Zhang, F. Phase transition and plastic deformation mechanisms induced by self-rotating grinding of GaN single crystals. *Int. J. Mach. Tools Manuf.* **2022**, *172*, 103827. [CrossRef]
2. Li, C.; Hu, Y.; Wei, Z.; Wu, C.; Peng, Y.; Zhang, F.; Geng, Y. Damage evolution and removal behaviors of GaN crystals involved in double-grits grinding. *Int. J. Extreme Manuf.* **2024**, *6*, 025103. [CrossRef]
3. Wu, Y.; Mu, D.; Huang, H. Deformation and removal of semiconductor and laser single crystals at extremely small scales. *Int. J. Extreme Manuf.* **2020**, *2*, 012006. [CrossRef]
4. Gao, S.; Song, W.; Huang, J.; Yang, X.; Kang, R. Effect of abrasives interference on deformation and material removal mechanism of single crystal YAG in abrasive machining. *Mater. Charact.* **2025**, *222*, 114804. [CrossRef]
5. Xu, D.; Wang, C.; Du, C.; Ding, F.; Hu, X. Physical and chemical characterization of the surface and removal process of silicon carbide ceramics by femtosecond laser processing. *Diam. Abras. Eng.* **2024**, *44*, 508–517. [CrossRef]

6. Li, C.; Wang, K.; Zakharov, O.; Cui, H.; Wu, M.; Zhao, T.; Yan, Y.; Geng, Y. Damage evolution mechanism and low-damage grinding technology of silicon carbide ceramics. *Int. J. Extreme Manuf.* **2025**, *7*, 022015. [CrossRef]
7. Xing, Y.; Xue, C.; Liu, Y.; Du, H.; Yip, W.S.; To, S. Freeform surfaces manufacturing of optical glass by ultrasonic vibration-assisted slow tool servo turning. *J. Mech. Work. Technol.* **2024**, *324*, 118271. [CrossRef]
8. Xu, P.; Sun, Y.; Zhang, G.; Kang, S.; Lu, W.; Sun, Y.; Zuo, D. Comparison of erosion resistance of hard and brittle materials processed by low-temperature micro-abrasive gas jet. *Diam. Abras. Eng.* **2024**, *44*, 665–674. [CrossRef]
9. Zhang, Z.; Shi, K.; Shi, Y.; Li, H.; Lu, D.; Kuang, Y.; Liu, J. Evolution mechanisms of the scratch-induced elastoplastic stress fields and crack damage in γ-TiAl alloys. *J. Mater. Res. Technol.* **2025**, *34*, 932–945. [CrossRef]
10. Song, Y.; Shi, K.; He, Z.; Wang, S.; Zhang, Z.; Shi, Y.; Huai, W. Wear characteristics evolution of corundum wheel and its influence on performance in creep feed grinding of nickel-based superalloy. *Wear* **2025**, *562*, 205649. [CrossRef]
11. Xiao, G.; Yang, Z.; Zhou, K.; He, Y.; Li, X. Significant improvement of machinability of Cf/SiC composites through matching laser scanning spacing and abrasive belt grain size. *Chin. J. Aeronaut.* **2025**, *38*, 103017. [CrossRef]
12. Qu, S.; Yao, P.; Gong, Y.; Yang, Y.; Chu, D.; Zhu, Q. Modelling and grinding characteristics of unidirectional C–SiCs. *Ceram. Int.* **2022**, *48*, 8314–8324. [CrossRef]
13. Wen, J.; Wang, Q.; Yu, A.; Wu, C. Removal mechanism of unidirectional Cf/SiC composites based on single diamond grit scratching. *Diam. Abras. Eng.* **2024**, *44*, 327–334. [CrossRef]
14. Li, C.; Wang, K.; Piao, Y.; Cui, H.; Zakharov, O.; Duan, Z.; Zhang, F.; Yan, Y.; Geng, Y. Surface micro-morphology model involved in grinding of GaN crystals driven by strain-rate and abrasive coupling effects. *Int. J. Mach. Tools Manuf.* **2024**, *201*, 104197. [CrossRef]
15. Li, C.; Li, X.; Wu, Y.; Zhang, F.; Huang, H. Deformation mechanism and force modelling of the grinding of YAG single crystals. *Int. J. Mach. Tools Manuf.* **2019**, *143*, 23–37. [CrossRef]
16. Li, C.; Zhang, F.; Meng, B.; Liu, L.; Rao, X. Material removal mechanism and grinding force modelling of ultrasonic vibration assisted grinding for SiC ceramics. *Ceram. Int.* **2017**, *43*, 2981–2993. [CrossRef]
17. Qu, S.; Yang, Y.; Yao, P.; Li, L.; Sun, Y.; Chu, D. Fiber reinforced ceramic matrix composites: From the controlled fabrication to precision machining. *Int. J. Extrem. Manuf.* **2025**, *7*, 062004. [CrossRef]
18. Li, C.; Liu, G.; Gao, C.; Yang, R.; Zakharov, O.; Hu, Y.; Yan, Y.; Geng, Y. Atomic-scale understanding of graphene oxide lubrication-assisted grinding of GaN crystals. *Int. J. Mech. Sci.* **2025**, *286*, 109934. [CrossRef]
19. Zhang, Z.; Yan, Y.; Wang, Z. The wear behavior and lapping performance of fixed abrasive pad for acidity and alkalinity of lapping fluid in lapping quartz glass. *Tribol. Int.* **2024**, *195*, 109615. [CrossRef]
20. Chen, H.; Wan, H.; Hong, B.; Hang, W.; Zhu, T.; Zhang, P.; Cao, X.; Xu, Q.; Wang, R.; Han, X.; et al. A novel liquid film shearing polishing technique for silicon carbide and its processing damage mechanisms. *Appl. Surf. Sci.* **2025**, *688*, 162317. [CrossRef]
21. Wang, L.; Hong, B.; Chen, H.; Qi, H.; Zhang, J.; Hang, W.; Han, Y.; Wang, J.; Ren, K.; Lyu, B. Enhancing tungsten machinability via laser pretreatment for abrasive particles-based shear rheological polishing. *Powder Technol.* **2025**, *455*, 120758. [CrossRef]
22. Guo, D. High-performance manufacturing. *Int. J. Extrem. Manuf.* **2024**, *6*, 060201.
23. Lv, B.; Lin, B.; Zhang, J.; Liu, C.; Wang, L.; Sui, T. Micro- and nano-scale spindle perpendicularity modulation method to enhance the quality of milled surfaces. *Int. J. Extreme Manuf.* **2024**, *6*, 065101. [CrossRef]
24. Xu, J.; Geier, N.; Shen, J.; Krishnaraj, V.; Samsudeensadham, S. A review on CFRP drilling: Fundamental mechanisms, damage issues, and approaches toward high-quality drilling. *J. Mater. Res. Technol.* **2023**, *24*, 9677–9707. [CrossRef]
25. Li, C.; Hu, Y.; Zhang, F.; Geng, Y.; Meng, B. Molecular dynamics simulation of laser assisted grinding of GaN crystals. *Int. J. Mech. Sci.* **2023**, *239*, 107856. [CrossRef]
26. Wang, J.; Fang, F.; An, H.; Wu, S.; Qi, Y.; Cai, Y.; Guo, G. Laser machining fundamentals: Micro, nano, atomic and close-to-atomic scales. *Int. J. Extrem. Manuf.* **2023**, *5*, 012005. [CrossRef]
27. Xu, N.; Kang, R.; Zhang, B.; Zhang, Y.; Wang, C.; Bao, Y.; Dong, Z. Improving fatigue properties of normal direction ultrasonic vibration assisted face grinding Inconel 718 by regulating machined surface integrity. *Int. J. Extreme Manuf.* **2024**, *6*, 035101. [CrossRef]
28. Li, Z.; Jiang, F.; Jiang, Z.; Tian, Z.; Qiu, T.; Zhang, T.; Wen, Q.; Lu, X.; Lu, J.; Huang, H. Energy beam-based direct and assisted polishing techniques for diamond: A review. *Int. J. Extreme Manuf.* **2024**, *6*, 012004. [CrossRef]
29. Zhang, J.; Zheng, Z.; Huang, K.; Lin, C.; Huang, W.; Chen, X.; Xiao, J.; Xu, J. Field-assisted machining of difficult-to-machine materials. *Int. J. Extreme Manuf.* **2024**, *6*, 032002. [CrossRef]
30. Desbiolles, B.X.E.; Bertsch, A.; Renaud, P. Ion beam etching redeposition for 3D multimaterial nanostructure manufacturing. *Microsyst. Nanoeng.* **2019**, *5*, 11. [CrossRef] [PubMed]
31. Li, C.; Piao, Y.; Meng, B.; Zhang, Y.; Li, L.; Zhang, F. Anisotropy dependence of material removal and deformation mechanisms during nanoscratch of gallium nitride single crystals on (0001) plane. *Appl. Surf. Sci.* **2022**, *578*, 152028. [CrossRef]

32. Zhang, H.; Zhang, J.; Gao, S.; Yan, X.; Chen, T. Effect of Shot-peening Residual Stress on Fatigue Life and Relaxation of TC4. *China Surf. Eng.* **2024**, *37*, 171–178. [CrossRef]
33. Gao, S.; Wang, H.; Huang, H.; Dong, Z.; Kang, R. Predictive models for the surface roughness and subsurface damage depth of semiconductor materials in precision grinding. *Int. J. Extreme Manuf.* **2025**, *7*, 035103. [CrossRef]
34. Zhang, Y.; Wang, Q.; Li, C.; Piao, Y.; Hou, N.; Hu, K. Characterization of surface and subsurface defects induced by abrasive machining of optical crystals using grazing incidence X-ray diffraction and molecular dynamics. *J. Adv. Res.* **2022**, *36*, 51–61. [CrossRef]
35. Qiang, B.; Shi, K.; Liu, N.; Ren, J.; Shi, Y. Integrating physics-informed recurrent Gaussian process regression into instance transfer for predicting tool wear in milling process. *J. Manuf. Syst.* **2023**, *68*, 42–55. [CrossRef]
36. Rossi, E.; Wheeler, J.M.; Sebastiani, M. High-speed nanoindentation mapping: A review of recent advances and applications. *Curr. Opin. Solid State Mater. Sci.* **2023**, *27*, 101107. [CrossRef]
37. Piao, Y.; Li, C.; Hu, Y.; Cui, H.; Luo, X.; Geng, Y.; Zhang, F. Nanoindentation induced anisotropy of deformation and damage behaviors of MgF2 crystals. *J. Mater. Res. Technol.* **2024**, *28*, 4615–4625. [CrossRef]
38. Li, C.; Piao, Y.; Zhang, F.; Zhang, Y.; Hu, Y.; Wang, Y. Understand anisotropy dependence of damage evolution and material removal during nanoscratch of MgF_2 single crystals. *Int. J. Extreme Manuf.* **2023**, *5*, 015101. [CrossRef]
39. Wang, J.; Yan, Y.; Cui, H.; Liu, S.; Li, C.; Geng, Y. Towards understanding the mechanisms of material removal and deformation in GaAs during nanomilling. *J. Mech. Work. Technol.* **2025**, *337*, 118712. [CrossRef]
40. Zhao, G.; Zhao, B.; Ding, W.; Xin, L.; Nian, Z.; Peng, J.; He, N.; Xu, J. Nontraditional energy-assisted mechanical machining of difficult-to-cut materials and components in aerospace community: A comparative analysis. *Int. J. Extreme Manuf.* **2024**, *6*, 022007. [CrossRef]
41. Lin, J.; Wu, M.; Liu, S.; Zhou, Y.; Gu, Y.; Zhou, X. Processing Mechanism and Experiment of Ultrasonic Vibration Assisted Cutting of SiCp/Al Composites. *China Surf. Eng.* **2024**, *37l*, 182–198. [CrossRef]
42. Malayath, G.; Mote, R.G. A review of cutting tools for ultra-precision machining. *Mach. Sci. Technol.* **2022**, *26*, 923–976. [CrossRef]
43. Hatefi, S.; Smith, F. Design and analysis of ultra-precision smart cutting tool for in-process force measurement and tool nanopositioning in ultra-high-precision single-point diamond turning. *Micromachines* **2023**, *14*, 1857. [CrossRef]
44. Zhao, B.; Wang, Y.; Peng, J.; Wang, X.; Ding, W.; Lei, X.; Wu, B.; Zhang, M.; Xu, J.; Zhang, L.; et al. Overcoming challenges: Advancements in cutting techniques for high strength-toughness alloys in aero-engines. *Int. J. Extrem. Manuf.* **2024**, *6*, 062012. [CrossRef]
45. Wang, K.; Li, C.; Wu, Y.; Liu, Y.; Zakharov, O.; Geng, Y.; Zhang, F. Nanoscale insight into brittle-to-ductile transition mechanism of 4H-SiC in single-grit nanogrinding. *Int. J. Mech. Sci.* **2025**, *302*, 110579. [CrossRef]
46. Jahnel, K.; Michels, R.; Wilhelm, D.P.; Grunwald, T.; Bergs, T. Investigation of Surface Integrity Induced by Ultra-Precision Grinding and Scratching of Glassy Carbon. *Micromachines* **2023**, *14*, 2240. [CrossRef]
47. Zhao, T.; Guo, L.; Gao, Q.; Wang, X.; Lyu, B.; Li, C. Modeling and Validation of Material Removal Based on Rheological Behavior Under Dynamic-Viscosity Nonlinear Coupling Effects. *Micromachines* **2025**, *16*, 572. [CrossRef]
48. Zhang, S.; Gao, Y.; Zhang, X.; Guo, Y. Influence of Diamond Wire Saw Processing Parameters on the Sawn Surface Characteristics of Silicon Nitride Ceramics. *Micromachines* **2023**, *14*, 1660. [CrossRef]
49. Zhang, H.; Niu, Y.; Jia, X.; Chu, S.; Niu, J. Longitudinal–Torsional Ultrasonic Grinding of GCr15: Development of Longitudinal–Torsional Ultrasonic System and Prediction of Surface Topography. *Micromachines* **2023**, *14*, 1626. [CrossRef] [PubMed]
50. Lin, H.; Zhou, M.; Wang, H.; Bai, S. Investigation of cutting force and the material removal mechanism in the ultrasonic vibration-assisted scratching of 2D-SiCf/SiC Composites. *Micromachines* **2023**, *14*, 1350. [CrossRef]
51. Yang, T.; Xiao, Y.; Hang, Y.; Wu, X.; Kong, W. Research on helical electrode electrochemical drilling assisted by anode vibration for jet micro-hole arrays on tube walls. *Micromachines* **2025**, *16*, 86. [CrossRef] [PubMed]
52. Cheng, X.; Cui, Z.; Chen, J.; Wang, Y.; Yang, L. Understanding the Processing Quality Problem for Cutting Ceramic Materials Using the Thermal-Controlled Fracture Method Induced by a Single-Surface Heat Source. *Micromachines* **2024**, *15*, 957. [CrossRef] [PubMed]
53. Liu, F.; Yu, A.; Wu, C.; Liang, S.Y. Process analysis and topography evaluation for monocrystalline silicon laser cutting-off. *Micromachines* **2023**, *14*, 1542. [CrossRef] [PubMed]
54. Qin, C.; Pan, J.; Guo, L.; Zhang, C.; Chen, W.; Hu, Z.; Mao, M. Experimental Study on Chemical–Mechanical Synergistic Preparation for Cemented Carbide Insert Cutting Edge. *Micromachines* **2024**, *15*, 17. [CrossRef] [PubMed]

Disclaimer/Publisher's Note: The statements, opinions and data contained in all publications are solely those of the individual author(s) and contributor(s) and not of MDPI and/or the editor(s). MDPI and/or the editor(s) disclaim responsibility for any injury to people or property resulting from any ideas, methods, instructions or products referred to in the content.

micromachines

Article

Modeling and Validation of Material Removal Based on Rheological Behavior Under Dynamic-Viscosity Nonlinear Coupling Effects

Tianchen Zhao [1], Luguang Guo [1,*], Qilong Gao [2], Xu Wang [2], Binghai Lyu [2] and Chen Li [3]

1 College of Mechanical Engineering, Quzhou University, Quzhou 324000, China; zhaotianchen1989@126.com
2 College of Mechanical Engineering, Zhejiang University of Technology, Hangzhou 310023, China; gao_qilong@163.com (Q.G.); wx382935877@163.com (X.W.); icewater7812@126.com (B.L.)
3 School of Mechatronics Engineering, Harbin Institute of Technology, Harbin 150001, China; hit_chenli@163.com
* Correspondence: glg7127@163.com

Abstract: Compliant rheological polishing advanced in facilitating the generation of smooth curved surfaces. However, the inherent energy dissipation of the medium during flow results in an uncontrollable material removal distribution. This study proposes utilizing the motion of the tool to regulate the distribution of physical fields within the computational domain, thereby controlling material removal. A film thickness model is developed based on fluid dynamics and tribology principles to examine the pressure and velocity distributions within the film. In conjunction with contact mechanics and metallography, a material removal model is formulated and then validated and refined by valid experiment, demonstrating a positive correlation between material removal rate and surface quality. Optimization experiments produced a curved surface with an *Ra* of 17.59 nm.

Keywords: non-Newtonian fluid; fluid dynamics; tribology principle; material removal

1. Introduction

Compliant fluid polishing technologies, such as Magnetorheological Finishing (MRF) and Shear-Thickening Polishing (STP), have garnered significant attention in recent years as non-traditional processing methods for high-precision surface machining [1–5]. Compared to traditional ultra-precision machining technology, these methods leverage the flow characteristics of liquid media to conform to the workpiece surface, enabling efficient material removal and facilitating the processing of complex curved surfaces and microstructures [6–12]. They are widely used in the final machining of optical components [13,14], aerospace components [15,16] and semiconductor devices [17,18].

However, due to the complex flow dynamics and shear stress distribution inherent in fluids, the material removal rate often exhibits substantial uncertainty [19], presenting a significant challenge for achieving high-precision surface machining. In compliant fluid polishing, the material removal distribution is primarily influenced by factors such as fluid viscosity, shear rate, abrasive concentration, and flow field distribution [20–22]. The nonlinear coupling of these factors makes it difficult to predict material removal distribution during polishing, complicating efforts to control uniformity in material removal.

Peng et al. analyzed the mechanism of inlet pressure drop during abrasive flow machining (AFM) processing using capillary flow and developed a constitutive model for high shear rate medium [23]. Kum et al. applied AFM to polish nozzle guide vanes, establishing a baseline for material removal distribution, which was later compensated by design modifications, reducing dimensional errors from 600 μm to 200 μm [24]. Building on this approach, Fu and Guo used AFM and STP to polish blades and bearing raceways, respectively, achieving high-precision material removal while maintaining excellent surface quality. After 400 cycles, Fu achieved a surface roughness *Ra* of 0.141 μm, and the standard deviation decreased from 0.0977 μm (without compensation) to 0.0291 μm [25]. Guo obtained a surface roughness of *Ra* 11.16 nm with a variance of 0.58 nm^2 after 90 min of polishing. Ke et al. employed shear thickening polishing (STP) to passivate cemented carbide inserts, achieving a reduction in cutting edge roughness from 118.01 nm to 8.13 nm within 10 min [26].

However, the compensation conditions for this approach are highly dependent on the workpiece geometry. Once the workpiece undergoes modification, a complete re-experimentation of the process is necessary, which limits the feasibility of accurately predicting the process. MRF provides a viable alternative: by calculating the relationship between the physical field distribution in the computational domain and the motion of the polishing tool, thereby enabling controlled material removal through carefully planned tool paths [27–29]. Zhu used a bonnet as the polishing tool and shear-thickening slurry as the medium, achieving a sub-micron bi-sinusoidal surface with a form error deviation of just 95 nm P–V [30,31]. Some researchers proposed magnetorheological fluid-assisted AFM methods, but these methods are essentially extensions of magnetorheological polishing and do not offer innovations to the underlying process [32,33]. Abrasive liquid jet polishing uses a jet of abrasive medium as a special polishing tool, which can also achieve high-precision material removal. However, its low processing efficiency, high nozzle wear, and rapid energy attenuation restrict further development of the technology [34,35].

This study examines the challenge of achieving deterministic material removal in compliant fluid polishing technologies, with a particular focus on shear-thickening fluids. It explores the physical field distribution within the computational domain of the polishing fluid assisted by the polishing tool and investigates the dynamic coupling mechanism between pressure and viscosity. By integrating experimental and numerical simulation methods, the results offer theoretical foundations and technical support for achieving high-precision control in compliant fluid polishing.

2. Principle of Dynamic-Viscous Shear-Thickening Polishing

Analysis of the shear thickening mechanism indicates that effective shear thickening polishing (STP) relies on generating a sufficiently high velocity gradient (shear rate) at the workpiece surface. However, under certain motion conditions, the high viscosity of the fluid leads to minimal slip between the slurry and the surface, resulting in negligible material removal even under high pressure. Additionally, the distribution of material removal is highly non-uniform.

To address this, a flexible tool movement is proposed to assist in STP processing. This movement actively creates a fluid shear zone at the workpiece surface, leading to localized rapid pressure and shear thickening. By leveraging the shear thickening effect of the slurry and its combined dynamic pressure lubrication effect, a high-viscosity, high-pressure polishing film is formed on the workpiece surface. This polishing film moves with the tool, generating relative motion between the tool and workpiece, thereby enabling ultra-precision processing of the surface.

The specific processing procedure is illustrated in Figure 1. The workpiece and polishing tool are immersed in the slurry. When the workpiece and tool are stationary in the fluid, the thickening phase exerts a lifting effect, causing the abrasive particles to be evenly dispersed in the slurry (Figure 1a). At this stage, there is no relative motion between the abrasive particles and the workpiece. As relative motion occurs between the workpiece and tool (Figure 1b), a velocity gradient is formed between them (Figure 1c). The shear thickening effect of the slurry causes a rapid increase in viscosity, and the thickening phase encapsulates the abrasive particles, driving them to scratch the workpiece surface and remove material. Figure 1d shows the force diagram of a single abrasive particle. The dynamic pressure p exerted on the particle is positively correlated with viscosity and velocity difference. High shear rates result in larger normal pressures and higher viscosities, which further increase the normal pressure. Under the combined action of these factors, the dynamic pressure p exerted by the polishing film is significantly higher than the p produced by the Newtonian slurry, thus enabling efficient material removal.

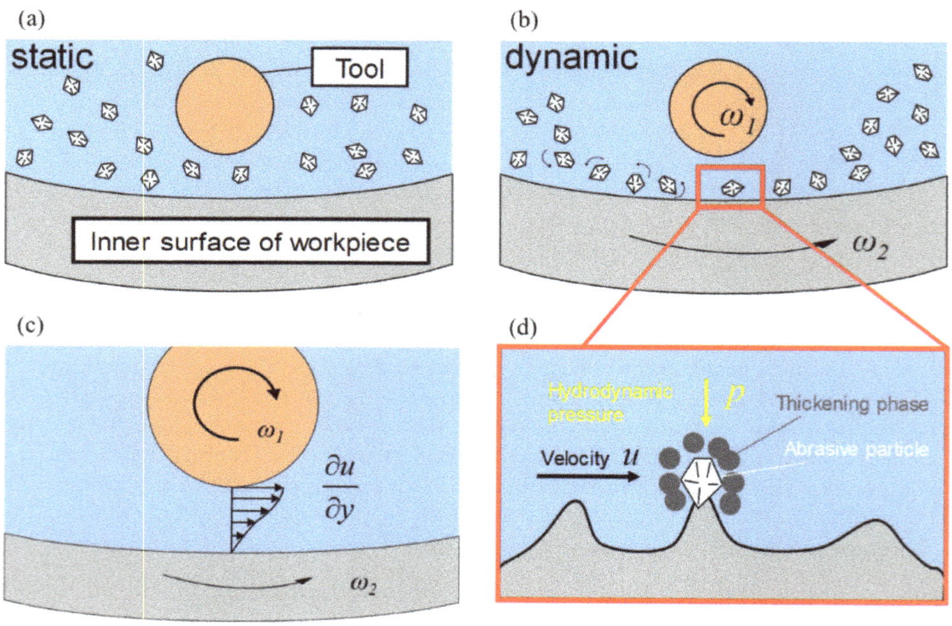

Figure 1. Principle of STP based on the dynamic-viscous pressure effect: (**a**) the slurry is static, (**b**) the slurry moves with the tool's rotation, (**c**) distribution of the velocity gradient, and (**d**) schematic of the force acting on the abrasives.

This method takes into account the coupling effect between pressure and viscosity, and is therefore referred to as the dynamic-viscous non-Newtonian polishing method (DVNNP). When the slurry is a shear-thickening fluid, this effect contributes to improved material removal. However, when the slurry is a shear-thinning fluid, the effectiveness should be evaluated based on the specific experimental conditions, comparing the influence of various factors on the magnitude of material removal.

Figure 2 illustrates the research approach in this paper. First, the processing model was simplified using Taylor expansion to define the effective processing area. The pressure and velocity partial differential equations were derived from the film lubrication equation, and numerical solutions were obtained using the Euler method. Material removal distribution was determined through metallographic analysis and contact mechanics and compared with experimental results for correction. The final model represents the material removal in non-Newtonian fluid polishing under dynamic-viscous pressure effects.

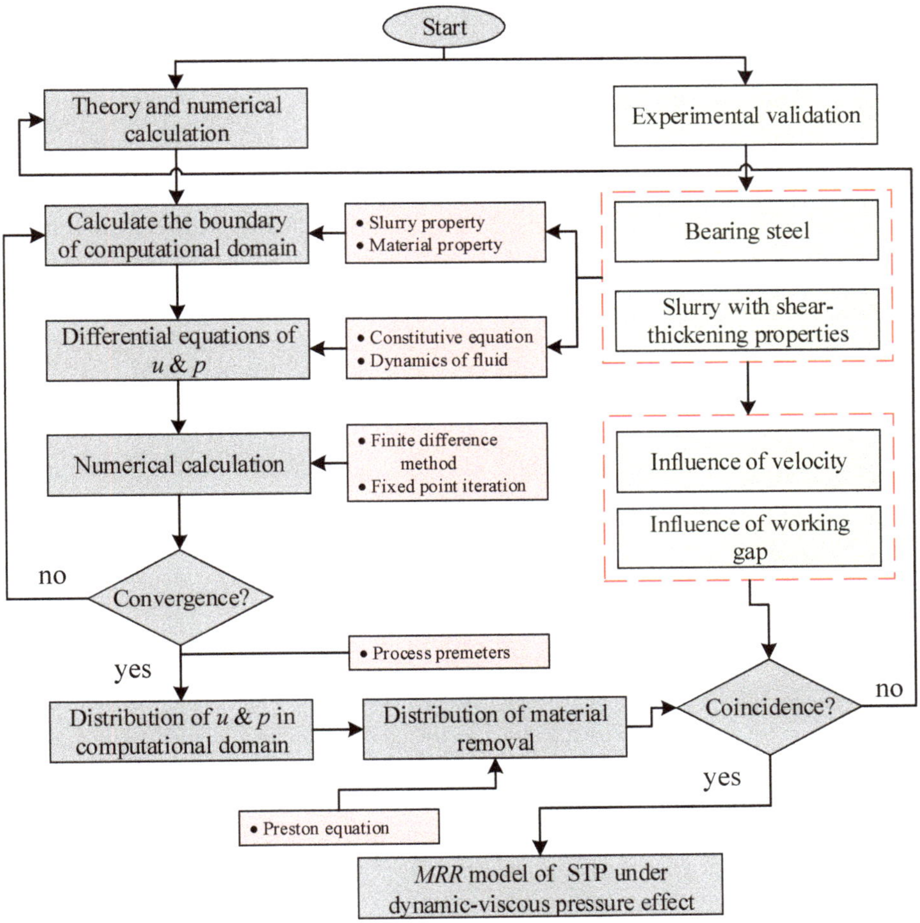

Figure 2. Flowchart of the research approach to material removal.

2.1. Assumptions

To ensure the logical rigor of the theoretical derivation, the following reasonable basic assumptions are necessary [36]:

(1) The dispersive phase and abrasives are uniformly distributed in the liquid. The transition from one control point to another is continuous, and the analysis can be conducted using infinitesimal methods. Discontinuities only occur at the boundaries, consistent with the continuous medium model.

(2) The properties of the material (including viscosity) may vary spatially, but such changes occur gradually, reflected in the spatial dependence of material properties in the continuous medium theory equations.

(3) Viscous forces dominate, while inertial forces can be neglected. Additionally, there is no relative motion between the thickened and solidified dispersive phase and the abrasives.

(4) The surface tension of the slurry is neglected, and due to the relatively short processing time in this experiment, temperature rise caused by friction and the evaporation of the liquid are ignored (according to a previous study) [37].

(5) The slurry is incompressible and isotropic; thus, the computational fluid dynamics is based on pressure solutions.

(6) Due to the high viscosity and the small thickness of the polishing film, pressure variation along the thickness direction is negligible.

All the analyses and conclusions in this paper are based on these assumptions.

2.2. *Rheological Performance*

To assess the performance retention of the slurry during the processing time, rheological curves before and after processing were measured for two sets. As shown in Figure 3, the rheological curves in the thickening region follow a power-law behavior, and the rheological performance is stable before and after processing.

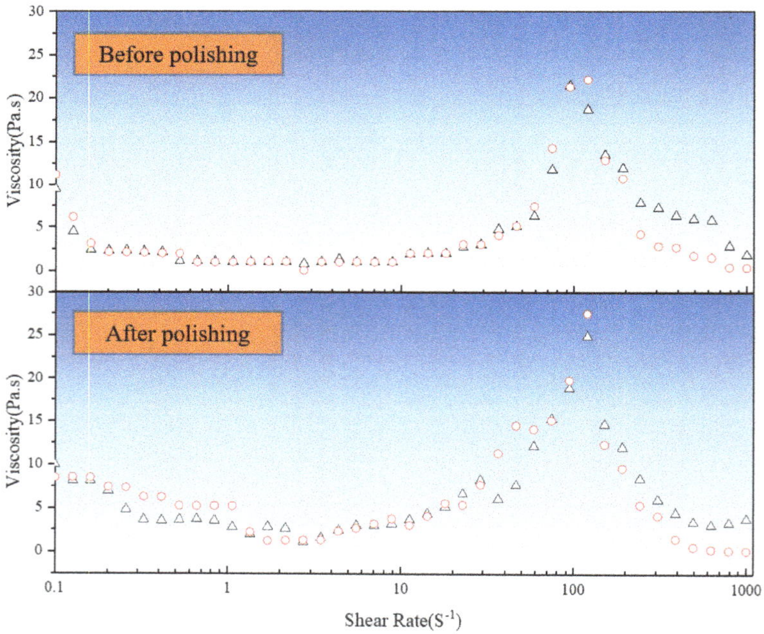

Figure 3. Comparison of rheological properties before and after polishing.

The fitted constitutive equation is given by the following equation:

$$\tau = \mu\dot{\gamma}^n = 0.008\dot{\gamma}^{2.6} \quad (1)$$

2.3. *Machining System*

In the research, the workpiece is the 32,234 cone roller bearing outer race, and the polishing tool selected is a flexible fiber to prevent damage to the workpiece. The experiment was conducted on the multi-axis bearing raceway polishing machine SRP370, as shown in Figure 4a. The polishing machine features bidirectional translation along the x and y axes, with the y-axis spindle positioned at the center of the workpiece. Both the polishing tool (Axis A) and the workpiece (Axis B) exhibit rotational motion and can achieve multi-axis linkage with the x and y spindles. The maximum rotational speed of Axis A is 300 rpm, and its rigidity ensures that the gap h along the raceway axis is uniform. Axis B has a maximum rotational speed of 100 rpm, while the maximum translation speed of the x and y spindles is 2000 mm/s. Figure 4b shows a schematic of the 32,234 polishing process, and the image in the upper-right corner of Figure 4a depicts the actual processing of the outer race, which is clamped in place by a fixture. Surface profiles and roughness were measured using the Taylor PGI810 surface profilometer (Taylor Hobson Ltd., Leicester, UK) and the Mar GD120 surface profilometer (Mahr GmbH, Göttingen, Germany).

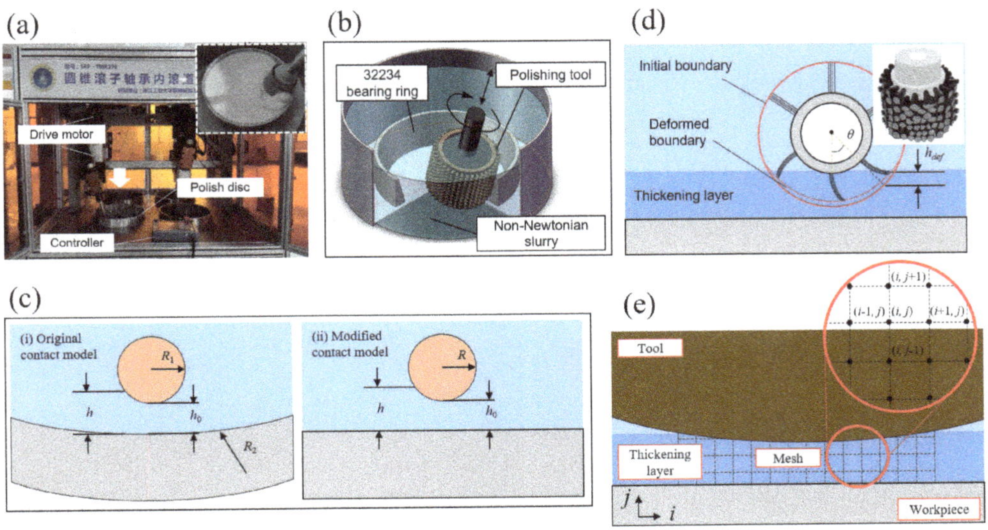

Figure 4. Overview of the manufacturing system: (**a**) prototype machine, (**b**) schematic of the polishing process, (**c**) computational domain boundaries, (**d**) simplification of the polishing model, and (**e**) illustration of mesh grid division.

3. Simulation and Analysis

3.1. Simplification of Geometric Models

In practical polishing processes, the polishing tool, which rotates around its axis, typically has a cylindrical cross-section, while the surface of the workpiece may have various curved shapes. Since the contact area width is much smaller than the curvature radius of the contact point, a geometric simplification of the contact surface is possible.

As shown in Figure 4c(i), R_1 represents the radius of the polishing head, and R_2 represents the curvature radius of the inner surface of the workpiece. The thickness of the polishing fluid film h can be expressed as follows:

$$h = h_0 + \left(R_1 - \sqrt{R_1^2 - x^2}\right) + (R_2 - \sqrt{R_2^2 - x^2}) \tag{2}$$

In the equation, h_0 represents the minimum gap between the tool and the surface. It is important to note that the size of the contact area is much smaller than both the radius of the tool and the curvature radius of the surface, i.e., $x \ll R_i$. Therefore, Equation (2) can be rewritten as follows:

$$h = h_0 + R_1(1 - \frac{x}{R_1}^2)^{\frac{1}{2}} + R_2(1 - \frac{x}{R_2}^2)^{\frac{1}{2}} \tag{3}$$

Expanding the term $(1 - \frac{x}{R_i}^2)^{\frac{1}{2}}$ using a Taylor series and neglecting higher-order small terms such as $\frac{x}{R_i}$, the following expression can be derived:

$$h = h_0 + \frac{x^2}{2R'} \tag{4}$$

In the equation, $R' = \frac{R_1 R_2}{R_2 - R_1}$ represents the equivalent radius of the geometric model, where $R_2 > R_1$.

Therefore, the geometric model is adjusted from the initial hyperboloid model shown in Figure 4c(i) to an equivalent cylindrical and planar contact model, as depicted in Figure 4c(ii).

The thickness h calculated above determines the initial position of the polishing tool. Due to the influence of the slurry's dynamic-viscous properties during the machining process, the boundary of the polishing tool, $B_{(x,y)}$, undergoes deformation (as shown in Figure 4d). The actual boundary of the polishing tool, $B_{(x,y)}$, can be expressed as follows:

$$B_{(x,y)} = B_{ini} + B_{def} \tag{5}$$

Here, B_{ini} represents the boundary of the tool when both the tool and the workpiece are stationary, as determined by the aforementioned thickness h. B_{def} represents the deformation caused by external forces. The deformation of the flexible fibers under shear force can be described by Equation (6):

$$w_B = \int_0^{l\sin\theta} \frac{pl^4}{8EI_y} d_x \tag{6}$$

In this equation, p is the load applied to the bristles, which depends on the spatial position x; l is the length of the bristles; E is the Young's modulus of the bristle fibers; and I_y is the moment of inertia of the bristles about the y-axis.

The physicochemical properties and structural dimensions of the polishing tool are shown in Table 1. According to previous studies [38], the maximum pressure at the center is on the order of 10^5 Pa. By inputting the pressure magnitude, structural dimensions, and material properties of the polishing tool into Equation (6), the elastic deformation of the polishing tool is calculated to be on the order of 0.09 mm. Considering the characteristic dimensions on the millimeter scale, this deformation is negligible. Therefore, the polishing tool can be considered to undergo no deformation.

Table 1. Properties and structural dimensions of polishing tool [39].

Tool Parameter	Value
Material	nylon fiber
Young's modulus	5 Gpa
Fiber diameter	0.5 mm
Fiber length	10 mm

3.2. Velocity Distribution

According to assumption (3), since inertial forces are neglected, the force balance equation shown in Figure 5 should have zero on the right-hand side:

$$\frac{\partial \sigma_{xx}}{\partial x} + \frac{\partial \sigma_{yx}}{\partial y} = 0 \tag{7}$$

In Equation (7), the fluid element is subjected only to normal pressure and viscous forces between the infinitesimal volumes. Therefore, the normal stress σ_{xx} is the negative of the pressure p (with compressive stress as negative and tensile stress as positive), while σ_{yx} represents the shear stress τ between adjacent fluid elements. Hence, Equation (7) can be rewritten as follows:

$$-\frac{\partial p}{\partial x}+\frac{\partial \tau}{\partial y}=0 \tag{8}$$

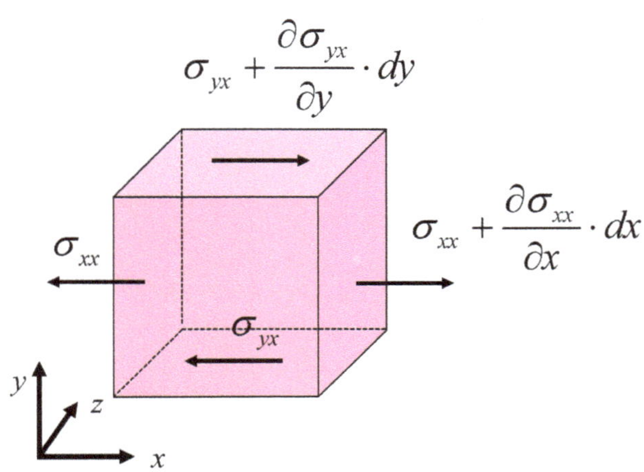

Figure 5. Schematic representation of the microscopic force acting on the slurry.

Based on assumption (6), it can be concluded that:

$$\frac{\partial p}{\partial y}=0 \tag{9}$$

Substituting the constitutive equation, Equation (1), into Equation (8):

$$\frac{1}{\mu_0}\frac{\partial p}{\partial x}=\frac{\partial}{\partial y}\left(\frac{\partial u}{\partial y}\right)^n \tag{10}$$

Considering that p is a function of x only, integrating both sides of Equation (10) gives the following equation:

$$\left(\frac{\partial u}{\partial y}\right)^n=\int \frac{1}{\mu_0}\frac{\partial p}{\partial x}d_y+C_1 \tag{11}$$

Integrating the above equation once more with respect to y:

$$u=\int\left(\frac{y}{\mu_0}\frac{\partial p}{\partial x}d_y+c_1\right)^{\frac{1}{n}}d_y+C_2 \tag{12}$$

This results in:

$$u=\frac{\mu_0}{\frac{\partial p}{\partial x}}\frac{n}{n+1}\left(\frac{y}{\mu_0}\frac{\partial p}{\partial x}d_y+c_1\right)^{\frac{n}{n+1}}+C_2 \tag{13}$$

Introducing the no-slip boundary condition at the wall, the velocity distribution boundary conditions at the workpiece surface and the tool surface are as follows:

$$\begin{cases} y = 0, u = 0 \\ y = h, u = U_t \end{cases} \tag{14}$$

Here, U_t is the velocity at the edge of the polishing tool.

Theoretically, substituting the boundary condition in Equation (14) into the velocity function, the constants C_1 and C_2 can be determined. However, an analytical solution for Equation (13) is difficult to obtain. In practice, although C_1 and C_2 are independent of y, they are functions of x. If an analytical solution for C_1 and C_2 in symbolic form cannot be found, it becomes challenging to obtain accurate results. Huang et al. [39,40] proposed a method for velocity separation. The theory is to decompose the velocity of a non-Newtonian fluid into two components: the shear velocity component u_1 (hereinafter referred to as "shear velocity"), which changes due to the velocity difference between the upper and lower boundaries, and the pressure velocity component u_2 (hereinafter referred to as "pressure velocity"), which changes due to variations in pressure.

The shear velocity is only related to the velocity difference between the upper and lower boundaries (i.e., U_t), so the partial derivative with respect to pressure is zero:

$$\frac{\partial p}{\partial x} = \frac{\partial}{\partial y}\left(g\left(\frac{\partial u_1}{\partial y}\right)\right) = 0 \tag{15}$$

In the above equations: g represents the function of the constitutive equation. Equation (15) indicates that the shear velocity distribution function is independent of y:

$$g\left(\frac{\partial u_1}{\partial y}\right) = \text{const}(x) \tag{16}$$

In this equation, const(x) represents a constant term that may depend on the function of x. Since the velocity gradient function of the shear rate is independent of y, its inverse function must also be a constant term that does not depend on y. Therefore, the integral of the velocity gradient can be expressed as follows:

$$u_1 = c_1 y + c_2 \tag{17}$$

The expression above must also satisfy the boundary conditions presented in Equation (14), allowing for the determination of the constant term:

$$u_1 = U_t \frac{y}{h} \tag{18}$$

To solve for the pressure velocity distribution, the constitutive equation is substituted into Equation (13) and integrated with respect to y. It is important to note that the pressure-driven flow exhibits symmetry:

$$\frac{\partial u_2}{\partial y} = \left(\frac{1}{\mu_0}\frac{\partial p}{\partial x}\left(y - \frac{h}{2}\right)\right)^{\frac{1}{n}} \tag{19}$$

Integrating the above equation and applying the wall boundary conditions allows the determination of the constant of integration, resulting in the velocity distribution:

$$u_2 = \frac{n}{n+1}\left(\frac{1}{\mu_0}\frac{\partial p}{\partial x}\right)^{\frac{1}{n}}\left(\left(y-\frac{h}{2}\right)^{1+\frac{1}{n}} - \left(\frac{h}{2}\right)^{1+\frac{1}{n}}\right) \tag{20}$$

By superimposing Equations (18) and (20), the velocity distribution of the slurry within the computational domain is obtained as follows:

$$u = u_1 + u_2 = U_t\frac{y}{h} + \frac{n}{n+1}\left(\frac{1}{\mu_0}\frac{\partial p}{\partial x}\right)^{\frac{1}{n}}\left(\left|y-\frac{h}{2}\right|^{1+\frac{1}{n}} - \left(\frac{h}{2}\right)^{1+\frac{1}{n}}\right) \tag{21}$$

Considering the nonlinearity introduced by the thickening exponent, the sign of Equation (21) needs to be adjusted to ensure the logical accuracy of the numerical calculations and avoid the appearance of imaginary components:

$$u = U_t\frac{y}{h} + sign\left(\frac{\partial p}{\partial x}\right)\frac{n}{n+1}\left|\frac{1}{\mu_0}\frac{\partial p}{\partial x}\right|^{\frac{1}{n}}\left(\left|y-\frac{h}{2}\right|^{1+\frac{1}{n}} - \left(\frac{h}{2}\right)^{1+\frac{1}{n}}\right) \tag{22}$$

Here, $sign(\frac{\partial p}{\partial x})$ represents the sign of $\frac{\partial p}{\partial x}$, indicating whether it is positive or negative.

3.3. Pressure Distribution

By substituting Equation (21) into the continuity equation and integrating it using the flow continuity condition, an equation with a separable partial derivative term for pressure can be obtained [41,42]. Upon separating the variables, Equation (23) is derived:

$$\frac{d_p}{d_x} = sign|h-\overline{h}|\frac{2\mu_0}{h^{2n+1}}U_t^n\left(2\left(2+\frac{1}{n}\right)\right)^n |h-\overline{h}|^n \tag{23}$$

Here, $sign\left(h-\overline{h}\right)$ represents the sign of $\left(h-\overline{h}\right)$, indicating whether it is positive or negative.

It is important to note that when the material of the polishing tool is changed, the deformation may become substantial enough that it can no longer be ignored. Consequently, the form of Equation (23) should be adjusted as follows:

$$\frac{d_p}{d_x} = f(p, w_B, x) = f(p, x) \tag{24}$$

Equation (24) indicates that the distribution of the boundary h is a function of pressure p and spatial position x, with pressure p (including velocity distribution u) being related to h. These two variables require iterative solving. To ensure both computational accuracy and convergence of the nonlinear equations in this case, the implicit RK4 method is used.

In this study, due to the near-rigid characteristics of the polishing tool, the right-hand side of Equations (22) and (23) does not contain differential terms; thus, the Euler finite difference method (Appendix A) is employed. According to the principles of tribology, positive pressure is generated only by flow through a shrinking gap, meaning that material removal occurs solely on the shrinking side. Therefore, the boundary of the computational domain is set to the left edge of the polishing tool's arc and the minimum gap, i.e., $x \in (-R, 0)$. Under the known geometric structure of the model, the boundary conditions can be expressed as follows:

$$p_{x=-R} = 0,\ p_{x=R} = 0 \tag{25}$$

The calculation begins with iteration from both the inlet and outlet. It is important to note that although there is only one unknown variable, p, two boundary conditions are required. This is because Equation (23) is derived through two integrations, and contains an $\overline{h}$, which is unknown. Therefore, $\overline{h}$ must be iteratively corrected during the calculation process, as shown in Figure 6.

The iteration starts with the boundary condition at the inlet, $p_{x=-R} = 0$, and an initial estimate for $\overline{h}$ is given (typically the arithmetic mean of the maximum and minimum film thicknesses). Then, $\overline{h}$ is adjusted iteratively to satisfy the outlet condition, $p_{x=R} = 0$. Since there is only one zero crossing, the bisection method is used for iteration. Additionally, due to the significant pressure variation near the center, the iteration step size is adjusted accordingly. The iteration proceeds until the calculated outlet pressure falls below 10^{-8}, with no upper limit on the number of iterations. The computational domain mesh is shown in Figure 4e, with a grid size of 0.01×0.01 mm. Once the pressure distribution is obtained, it is substituted into Equation (22) to determine the velocity distribution.

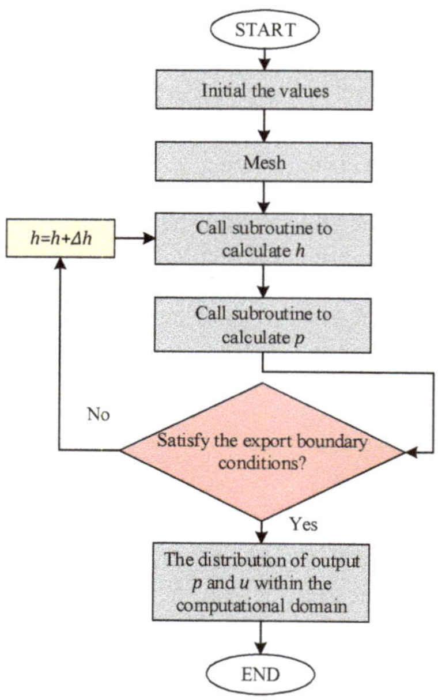

Figure 6. Flowchart of the proposed CFD.

3.4. *Removal Model of Single Abrasive*

After determining the velocity and pressure distributions within the computational domain, it is essential to calculate the number of effective abrasive particles involved in the process. Based on assumption (3), the abrasive particles are uniformly distributed, as illustrated in Figure 7a. Since the experimental system measures the mass concentration of the abrasive particles, it is necessary to establish a relationship between mass and volume concentrations [43] as follows:

$$C_v = \frac{\rho_m C_m}{\rho_m C_m + \rho_a (1 - C_m)} \tag{26}$$

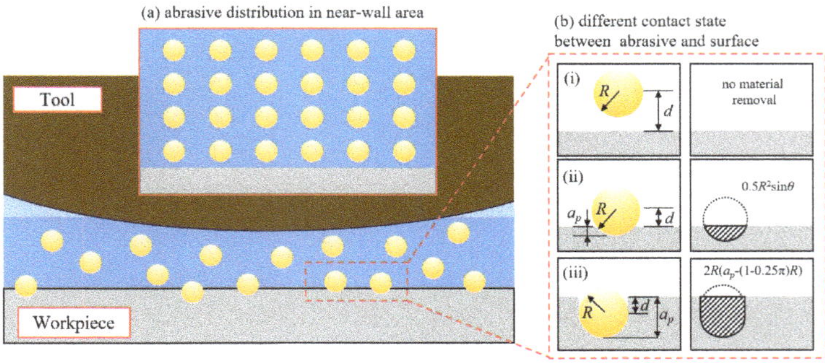

Figure 7. State of abrasives in the near-wall region: (**a**) abrasive distribution assumption and (**b**) abrasive-surface contact state, where: (**i**) $d > R$ & $a_p < 0$, (**ii**) $d < R$ & $a_p > 0$, (**iii**) $a_p > R$.

Assuming the abrasive particles are spherical, the number of effective particles per unit area, N_a, is expressed as follows:

$$N_a = \frac{3C_v}{4\pi R^3} d_s \tag{27}$$

In the abrasive polishing process, the contact between the particles and the surface can be classified into three types, as illustrated in Figure 7. When the polishing fluid pressure is insufficient to press the particles into the material surface ($d > R$), no material removal occurs. When the pressure is sufficient to press the particles to a certain depth ($d < R$), the projected area of the volume removed by a single particle in the direction of motion is $\frac{1}{2}R^2\sin\theta$, and the removal efficiency is $\frac{1}{2}R^2\sin\theta \cdot u$. When the pressure is high enough to embed the geometric center of the particle below the surface ($a_p > R$), the projected area of the volume removed by a single particle in the direction of motion is $2R(a_p - (1 - \frac{\pi}{4})R)$, and the removal efficiency is $2R(a_p - (1 - \frac{\pi}{4})R) \cdot u$. The depth of particle penetration, a_p, is determined based on contact mechanics.

$$a_p = \sqrt{\frac{2F_n}{\pi \sigma_s \tan^2 \alpha_c}} \tag{28}$$

3.5. Material Removal Model

Material removal function can be expressed as Equation (29), based on the analyses above:

$$V_{MRR} = \int_0^S \int_0^T \frac{3upC_v}{4\pi \sigma_s R tan\alpha_c} d_t d_s \tag{29}$$

By substituting the equations for the physical quantities within the computational domain, the complete material removal distribution function is obtained as follows:

$$V_{MRR} = \int_0^T \int \frac{3upC_v}{4\pi \sigma_s R tan\alpha_c} \int \frac{2\mu_0}{h^{2n+1}} U^n \left(2\left(2 + \frac{1}{n}\right)\right)^n \left(h - \bar{h}\right)^n d_t d_x (U_t \frac{y}{h} + \frac{n}{1+n}\left(\frac{1}{\mu_0}\frac{\partial p}{\partial x}\right)^{\frac{1}{n}})(|y - \frac{h}{2}|^{\frac{1}{n}+1} - \left(\frac{h}{2}\right)^{\frac{1}{n}+1}))d_s \tag{30}$$

4. Validation and Analysis

4.1. Valid Experiment Design

The experiment aims to investigate the effects of different relative velocities (u) and gap distances (h) on material removal, thereby validating the accuracy of the material removal model. A series of single-factor experiments were conducted to assess the uniformity of material removal under varying u and h, with a polishing time of 60 min. The levels of the other process parameters are provided in Table 2.

Table 2. Process parameters for validation experiments.

Processing Parameters	Value
Rotation speed of workpiece (rpm)	10, 25, 40
Inner diameter of workpiece (mm)	252~298
Rotation speed of tool (rpm)	100
Diameter of tool (mm)	60
Gap h (mm)	0.5, 1, 2

Due to the large mass of the workpiece and the relatively small amount of material removal, a precision balance cannot simultaneously satisfy both measurement requirements. To accurately quantify material removal during the processing cycle, three evenly spaced scratches were made axially on the outer bearing (32,234) raceway (Figure 8a), corresponding to inner diameters of 285 mm, 270 mm, and 255 mm. The yield strength of the material without heat treatment is 400 MPa. The material removal rate during this period was determined by measuring the change in distance, Δh, between the bottom of the scratches and the lower top surfaces before and after processing (Figure 8b).

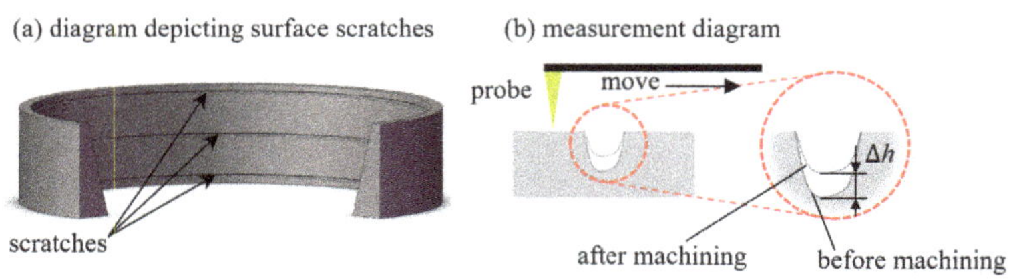

Figure 8. Scratch distribution and measurement: (**a**) scratch schematic and (**b**) scratch measurement.

4.2. Validation Results and Discussion

Figures 9–13 show the pressure and velocity distributions in the effective processing region under different relative speeds and gap sizes. Due to Equation (15) ($\frac{\partial p}{\partial y} = 0$), the pressure distribution is presented as a 2D line plot, with green, purple, and red regions representing the pressure distributions at the top, middle, and bottom of the raceway, respectively. The velocity distribution is shown as a contour plot. Unless otherwise specified, the data refers to the position at the top of the raceway.

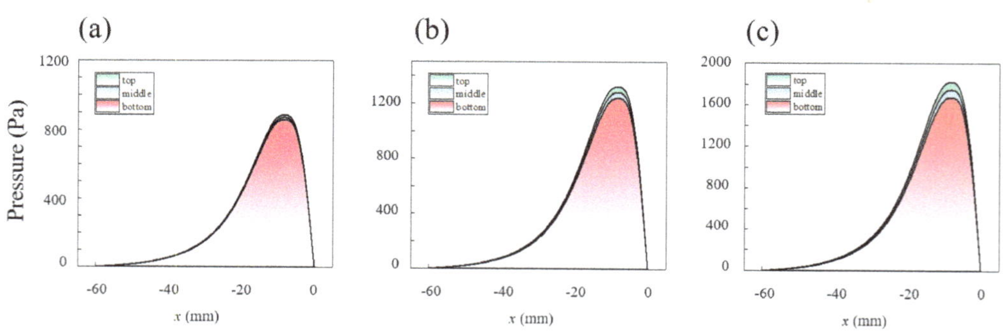

Figure 9. Effect of different velocities on the pressure distribution of the slurry within the computational domain: (**a**) 10 rpm, (**b**) 25 rpm, and (**c**) 40 rpm.

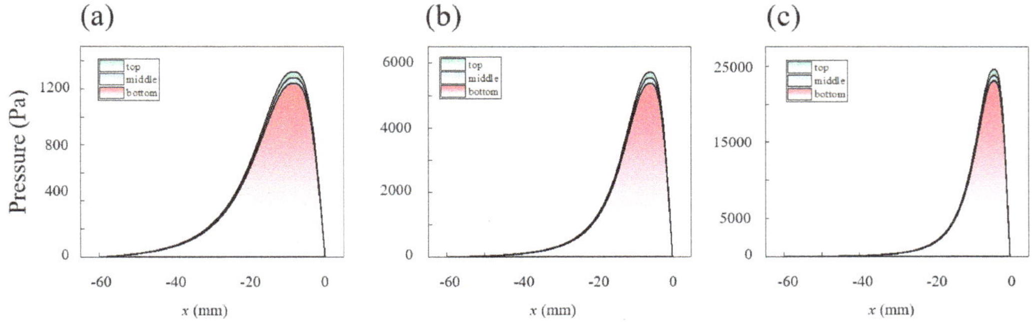

Figure 10. Effect of gap on the pressure distribution of the slurry within the computational domain: (**a**) 2 mm, (**b**) 1 mm, and (**c**) 0.5 mm.

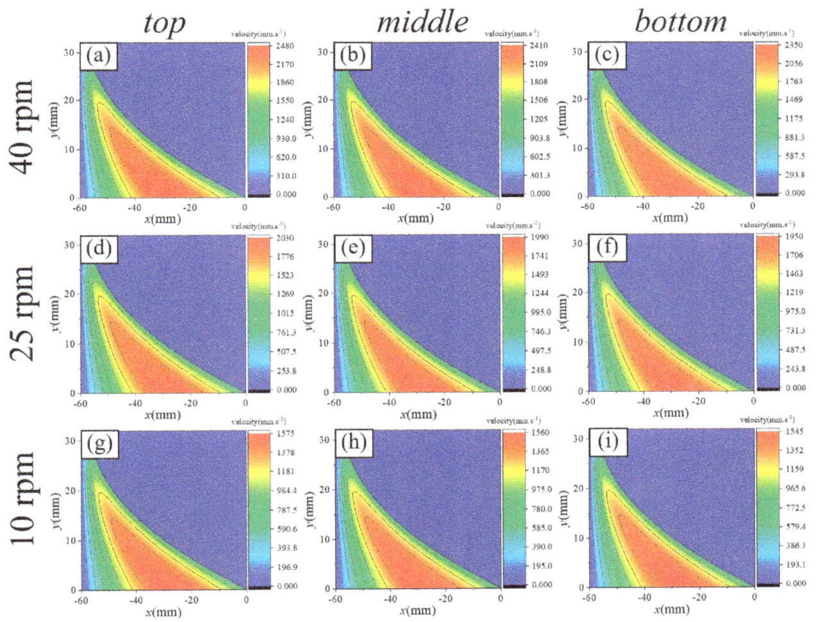

Figure 11. Effect of U on the velocity distribution of the slurry within the computational domain. Classified by speed: (**a–c**) with a rotation velocity of 40 rpm, (**d–f**) with a rotation velocity of 25 rpm, (**g–i**) with a rotation velocity of 10 rpm. Classified by position: (**a**,**d**,**g**) on the top of the raceway, (**b**,**e**,**h**) on the top of the raceway, (**c**,**f**,**i**) on the top of the raceway.

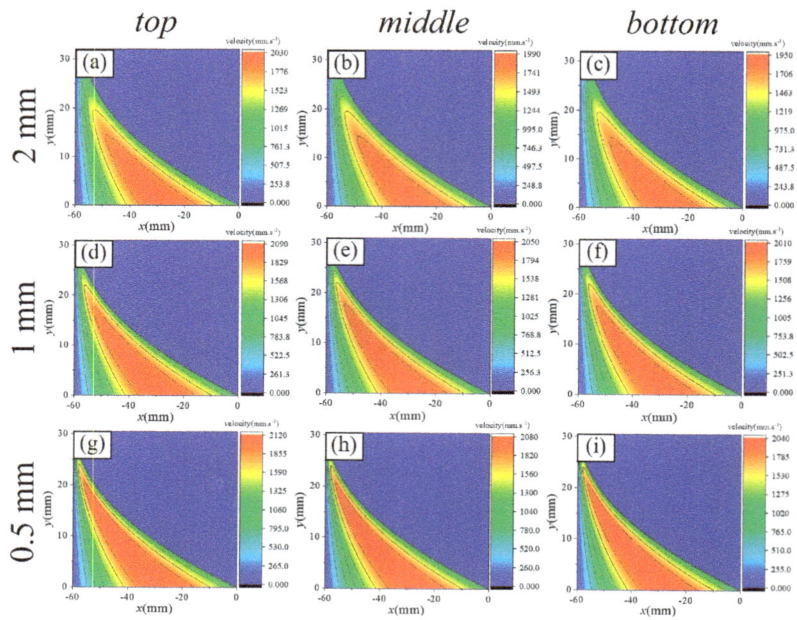

Figure 12. Effect of h on the velocity distribution of the slurry within the computational domain. Classified by speed: (**a–c**) with a rotation velocity of 40 rpm, (**d–f**) with a rotation velocity of 25 rpm, (**g–i**) with a rotation velocity of 10 rpm. Classified by position: (**a**,**d**,**g**) on the top of the raceway, (**b**,**e**,**h**) on the top of the raceway, (**c**,**f**,**i**) on the top of the raceway.

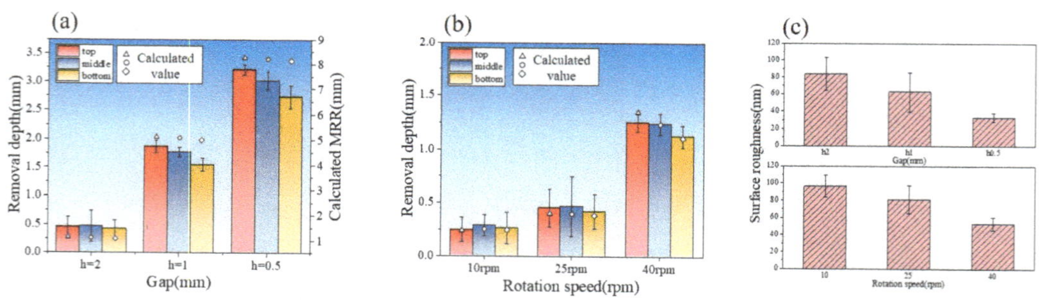

Figure 13. Validation result: (**a**,**b**) comparison of material removal between validation results and theoretical calculations at varying gaps and speeds and (**c**) evaluation of surface roughness under different processing conditions.

4.2.1. Analysis of Simulation Results

Pressure Distribution Law

The pressure variation in the slurry along the flow direction within the computational domain is shown in Figure 9 ($h_0 = 2$ mm) and Figure 10 ($\omega = 25$ rpm). The pressure jump occurs near the minimum gap (pressure outlet). In comparison to Newtonian fluids or shear-thinning fluids, the peak shape of shear-thickening fluids is much steeper, indicating that the viscosity change enhances the amplitude of the pressure variation [40].

Furthermore, the relationship between pressure distribution, rotational speed, and gap can be summarized as follows:

There is a pressure difference between the upper and lower parts of the raceway, caused by the velocity differences induced by the raceway structure. The pressure is highest at the inner diameter of the raceway top, which results in the pressure distribution being highest at the top, although the difference is small.

The pressure peak difference between different areas and the relative speed are proportional. When the raceway rotational speed is 10 rpm, the maximum pressure is 885.9 Pa, with a pressure difference of 26.9 Pa between the top and bottom. In contrast, when the raceway rotational speed is 40 rpm, the maximum pressure reaches 1821.7 Pa, with a pressure difference of 146.4 Pa between the top and bottom.

As the gap decreases, the pressure peak and the pressure difference at different areas along the raceway increase, and the pressure peak shifts away from the pressure outlet. When the gap is 2 mm, the pressure peak is 1322.6 Pa, occurring 0.84 mm from the outlet, with a pressure difference of 80.7 Pa between the top and bottom. When the gap is 0.5 mm, the pressure peak increases to 24,569.6 Pa, occurring 0.42 mm from the outlet, with a pressure difference of 1498.8 Pa between the top and bottom. The pressure difference at this gap even exceeds the peak pressure at a 2 mm gap.

Velocity Distribution Law

The variation in velocity along the flow direction within the computational domain is depicted in Figure 11 ($h_0 = 2$ mm) and Figure 12 ($\omega = 25$ rpm). Figure 11a–c, Figure 11d–f, and Figure 11g–i illustrate the velocity distribution of the slurry within the computational domain at ring rotational speeds of 40 rpm, 25 rpm, and 10 rpm, respectively. Figure 12a–c, Figure 12d–f, and Figure 12g–i show the velocity distribution of the slurry for initial minimum gaps of 2 mm, 1 mm, and 0.5 mm, respectively. The peak velocity of the slurry is observed not at the wall surface, but in the central region of the flow. For material removal calculations, velocity values near the wall must be extracted. Furthermore, the relationship between the velocity distribution, rotational speed, and gap can be summarized as follows:

(1) At the same rotational speed, varying raceway radii across different regions result in different relative velocities. However, these differences are minimal, with the maximum variation being only 130 mm/s.

(2) The ring rotational speed has the most significant impact on relative velocity. As the ring speed decreases from 40 rpm to 10 rpm, the peak velocity of the slurry decreases by 955 mm/s. Notably, the peak velocity does not occur at the wall but in the central region of the flow. For material removal calculations, velocity values near the wall are necessary.

(3) As the gap decreases, the peak velocity increases; however, this increase is much smaller compared to the change in pressure, as the increase in viscosity hinders slurry motion. When the gap is reduced from 2 mm to 0.5 mm, the peak velocity increases from 2030 mm/s to 2120 mm/s.

In summary, the gap exerts a substantial influence on the pressure distribution of the slurry, with its effect on pressure being an order of magnitude (10^1 times) greater than that of relative velocity. Moreover, higher rotational speeds lead to increased slurry velocities near the wall. Therefore, processing conditions characterized by small gaps and high relative velocities optimize material removal efficiency.

4.2.2. Validation Result

Figure 13 presents a comparison between the experimental results and theoretical calculations of material removal efficiency under different process parameters. Figure 13a compares the results of the removal model calculations and experiments at $\omega = 25$ rpm for different gap sizes. When the minimum gap is 2 mm, the removal thickness of the inner raceway ranges from 0.417 to 0.453 μm, with a theoretical model median of 1.103 μm. At a gap of 1 mm, the removal thickness ranges from 1.549 to 1.872 μm, with a model median of 5.082 μm. For a gap of 0.5 mm, the removal thickness ranges from 2.732 to 3.209 μm, with a model median of 8.214 μm. Due to the assumptions made in the model, deviations from actual processing conditions are inevitable, necessitating model correction. As shown in Figure 13a, the deviation between the theoretical model and experimental results is linear, allowing for the application of a linear correction factor, k. The value of k is determined as the arithmetic mean of all deviations, $k = 0.361$. After correction, the maximum error occurs at the small gap of $h_0 = 0.5$ mm, at approximately 16.3%, with an average error of about 12.5%.

The comparison between the corrected model's calculated values and theoretical values at different rotational speeds was also examined. Figure 13b compares the model and experimental values at $h_0 = 2$ mm for different rotational speeds. When the ring speed is 10 rpm, the removal thickness of the inner raceway ranges from 0.247 to 0.287 μm, with a model median of 0.237 μm. At 25 rpm, the removal thickness ranges from 0.417 to 0.453 μm, with a model median of 0.397 μm. At 40 rpm, the removal thickness ranges from 1.124 to 1.249 μm, with a model median of 1.346 μm. At higher rotational speeds, the model shows a maximum error of approximately 14.0%, with an average error of about 11.1%.

In summary, the corrected model exhibits a maximum error of no more than 17% and an average error of no more than 13%. The datum is well-distributed and accurate, making it a reliable tool for material removal prediction.

Figure 13c presents a comparison of surface roughness (*Ra*) after polishing under various conditions. As the roughness values at the top, middle, and bottom showed minimal differences, the figure displays the overall average roughness across all positions. The roughness variation follows the same trend as material removal efficiency: higher rotational speeds and smaller gaps lead to the lowest roughness. Therefore, the model-predicted material removal efficiency can serve as an effective indicator for optimizing the process parameters that influence surface quality.

4.2.3. Optimal Experiment

Based on the analysis and experimental results, 0.5 mm and 40 rpm were identified as the optimal process parameters, with other variables held constant. Figure 14 illustrates the evolution of surface roughness in terms of *Ra* over time under these optimal conditions. After 30 min of polishing, the surface roughness decreased to 27.74 nm, outperforming all results obtained under different parameter settings in the verification tests. After 60 min of polishing, the surface roughness further decreased to 17.59 nm, demonstrating high consistency in surface quality, with the roughness variance of only 4.42 nm^2.

Figure 15 provides a comparison of the true picture before and after polishing. Before polishing (Figure 15a), the raceway surface, following precision grinding, exhibited a microstructure with groove-like stripes, and the surface showed a blurred reflective quality. In contrast, after polishing (Figure 15b), the raceway surface was free from visible machining marks, presenting a smooth, bright, mirror-like finish.

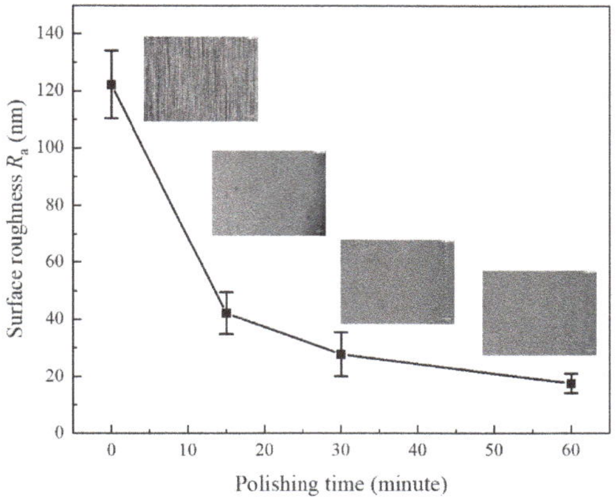

Figure 14. Variation in surface roughness with polishing time.

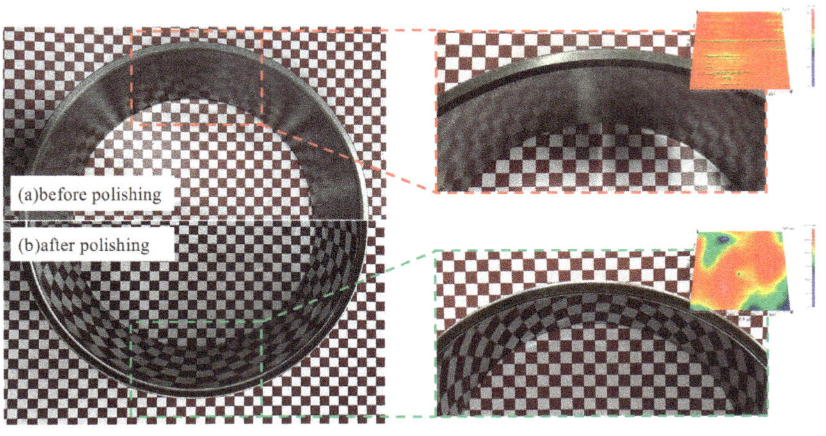

Figure 15. Comparison of surface quality of 32,234 raceways before and after STP, where (**a**) after grinding and (**b**) after STP processing.

5. Conclusions

This paper presents an advanced shear-thickening polishing method that applies normal pressure to the workpiece surface assisted by the motion of a tool. A material removal distribution model is developed, with the processing target being the raceway of a large 32,234 bearing outer ring. The model is refined through valid experiments. The main research contents and conclusions are as follows:

(1) The rheological properties of the slurry were measured, and reasonable assumptions were established based on machining conditions and relevant theory.

(2) Distribution functions for pressure and velocity within the processing area were derived. The velocity distribution within the computational domain was obtained using velocity decomposition, then the pressure distribution was computed using the finite difference method.

(3) A single abrasive removal function was established based on contact mechanics, and a material removal distribution function was derived using metallographic analysis.

(4) Valid experiments were carried out under different process parameter levels. The results indicated that the average error of the modified theoretical model was approxi-

mately 12%, with a maximum error of 16.3%. The model demonstrated high accuracy and high reliability.

(5) The surface roughness and material removal rate exhibited similar trends, allowing the removal rate to serve as an indicator for identifying the optimal process parameters for achieving the best surface quality. The optimization experiment yielded a surface with a roughness of *Ra* 17.59 nm and a variance of 4.42 nm^2, demonstrating good processing consistency.

This research presented in this paper aims to provide reliable material removal predictions for the rheological flexible polishing process, offering significant engineering application value.

Author Contributions: Conceptualization: T.Z.; data curation: Q.G. and X.W.; validation experiments: T.Z.; measurement: L.G.; writing—original draft: L.G. and C.L.; review: B.L. and C.L. All authors have read and agreed to the published version of the manuscript.

Funding: This research was funded by the National Natural Science Foundation of China (grant number 52375420, 52305483 and 51805485), the National Natural Science Foundation of Zhejiang province (grant number QZSZ25E050002, QZQN25E050002, and LMS25E050009). And The APC was funded by the Science and Technology Plan Project of Quzhou (grant number. 2023 K207).

Data Availability Statement: The datasets generated and/or analyzed during the current study are not publicly available due to the laboratory policy but are available from the corresponding author on reasonable request.

Conflicts of Interest: We declare that we have no financial and personal relationships with other people or organizations that can inappropriately influence our study, and there is no professional or other personal interest of any nature or kind in any product, service, and/or company that could be construed as influencing the position presented in, or the review of, the manuscript entitled.

Nomenclature

Nomenclature		a_p	depth of particle penetration
B	actual boundary	B_{ini}	boundary without deformance
B_{def}	deformation of the tool	C_v	volume concentrations
E	tool's Young's modulus	F	force on the tool
F_n	normal component of F	F_t	tangential component of F
h	gap size	h_0	minimum film thickness
I_y	inertia moment	k	correction factor
l	fiber length	n	thickening index
N_a	number of active abrasive	p	slurry pressure
R	equivalent radius	R_1	tool radius
R_2	workpiece radius	T	residence time
U_h	upper boundary rate	U_0	lower boundary rate
u	rate component in the x-direction	v	rate component in the y-direction
V_{MRR}	material removal rate	w_B	nylon deflection
δ	film thickness	θ	the central angle corresponding to the interval between different clusters of bristles
μ	viscosity	μ^*	apparent viscosity
ρ	density		

Appendix A. Euler's Difference Method and Its Implementation

The differential equation is expressed in the form, as shown in Equation (A1):

$$y' = f(x, y) \tag{A1}$$

The Euler forward finite difference method approximates the differential value using the curvature at a given point, which can be expressed as follows:

$$y'(x_{n+1}) = \frac{y(x_{n+1}) - y(x_n)}{x_{n+1} - x_n} \tag{A2}$$

where $y'(x_{n+1})$ is the differential value of the function at x_{n+1},

$y(x_{n+1})$ is the value of the function at x_{n+1},

$y(x_n)$ is the value of the function at x_n,

$x_{n+1} - x_n$ is the step size, denoted as Δx in subsequent expressions.

Therefore, the function value at the next grid point can be expressed as follows:

$$y(x_{n+1}) = y(x_n) + (x_{n+1} - x_n)f(x, y) \tag{A3}$$

Taking the calculation of pressure distribution as an example: discretizing Equation (A3) using the finite difference method and substituting the initial pressure values, the pressure distribution can be calculated using Equation (A4) as follows:

$$p_{i+1} = p_i + \Delta x \cdot f(x, y) \tag{A4}$$

Performing a Taylor series expansion of $y(x_{n+1})$ around x_n, the function can be expressed as follows:

$$y(x_{n+1}) = y(x_n) + \Delta x \cdot y' + o(\Delta x) \tag{A5}$$

Thus, the accuracy of the Euler forward finite difference method is of the order Δx, making it less suitable for nonlinear, non-homogeneous, or more complex differential equations. In such cases, improved Euler methods or other higher-order methods should be used. The code for the Euler forward finite difference method is not presented here due to its simplicity.

References

1. Wang, L.; Wu, M.; Chen, H.; Hang, W.; Wang, X.; Han, Y.; Chen, H.; Chen, P.; Beri, T.H.; Luo, L.; et al. Damage evolution and plastic deformation mechanism of passivation layer during shear rheological polishing of polycrystalline tungsten. *J. Mater. Res. Technol.-JMRT* **2024**, *28*, 1584–1596. [CrossRef]
2. Yao, W.; Chu, Q.; Lyu, B.; Wang, C.; Shao, Q.; Feng, M.; Wu, Z. Modeling of material removal based on multi-scale contact in cylindrical polishing. *Int. J. Mech. Sci.* **2022**, *223*, 19. [CrossRef]
3. Wang, J.; Tang, Z.; Goel, S.; Zhou, Y.; Dai, Y.; Wang, J.; He, Q.; Yuan, J.; Lyu, B. Mechanism of material removal in tungsten carbide-cobalt alloy during chemistry enhanced shear thickening polishing. *J. Mater. Res. Technol.-JMRT* **2023**, *25*, 6865–6879. [CrossRef]
4. Wang, Z.X.; Ye, L.Z.; Zhu, X.J.; Liu, Y.; Chuai, S.D.; Lv, B.Y.; Wang, D. Analysis of flow field characteristics of silicon carbide CMP under ultrasonic action. *Diam. Abras. Eng.* **2025**, *45*, 102–112.
5. Cheng, F.; Wang, Z.; Zun, R.; Wang, Y.G.; Peng, Y.; Zhang, T.Y.; Zhao, D.; Fan, C. Study on dispersion of abrasive particles in electro Fenton CMP slurry and design of green polishing fluid in neutral environment. *Diam. Abras. Eng.* **2025**, *45*, 113–121.
6. Lv, S.C.; Liu, Y.; Sun, X.W.; Dong, Z.X.; Yang, H.R.; Zhang, W.F. Numerical simulation of the influence of cutting parameters on the cutting process of ZrO_2 ceramics. *Diam. Abras. Eng.* **2024**, *44*, 769–780.
7. Huang, S.Q.; Li, X.L.; Zhao, Y.T.; Sun, Q.; Huang, H. A novel lapping process for single-crystal sapphire using hybrid nanoparticle suspensions. *Int. J. Mech. Sci.* **2021**, *191*, 12. [CrossRef]

8. Huang, S.Q.; Li, X.L.; Mu, D.K.; Cui, C.C.; Huang, H.; Huang, H. Polishing performance and mechanism of a water-based nanosuspension using diamond particles and GO nanosheets as additives. *Tribol. Int.* **2021**, *164*, 13. [CrossRef]
9. Qu, S.; Wei, C.; Yang, Y.; Yao, P.; Chu, D.; Gong, Y.; Zhao, D.; Zhang, X. Grinding mechanism and surface quality evaluation strategy of single crystal 4H-SiC. *Tribol. Int.* **2024**, *194*, 11. [CrossRef]
10. Li, C.; Wang, K.; Piao, Y.; Cui, H.; Zakharov, O.; Duan, Z.; Zhang, F.; Yan, Y.; Geng, Y. Surface micro-morphology model involved in grinding of GaN crystals driven by strain-rate and abrasive coupling effects. *Int. J. Mach. Tools Manuf.* **2024**, *201*, 25. [CrossRef]
11. Li, C.; Wang, K.; Zakharov, O.; Cui, H.; Wu, M.; Zhao, T.; Yan, Y.; Geng, Y. Damage evolution mechanism and low-damage grinding technology of silicon carbide ceramics. *Int. J. Extrem. Manuf.* **2025**, *7*, 36. [CrossRef]
12. Li, C.; Liu, G.; Gao, C.; Yang, R.; Zakharov, O.; Hu, Y.; Yan, Y.; Geng, Y. Atomic-scale understanding of graphene oxide lubrication-assisted grinding of GaN crystals. *Int. J. Mech. Sci.* **2025**, *286*, 19. [CrossRef]
13. Lin, B.; Jiang, X.M.; Cao, Z.C.; Huang, T. Development and theoretical analysis of novel center-inlet computer-controlled polishing process for high-efficiency polishing of optical surfaces. *Robot. Comput.-Integr. Manuf.* **2019**, *59*, 1–12. [CrossRef]
14. Su, X.; Ji, P.; Liu, K.; Walker, D.; Yu, G.; Li, H.; Li, D.; Wang, B. Combined processing chain for freeform optics based on atmospheric pressure plasma processing and bonnet polishing. *Opt. Express* **2019**, *27*, 17979–17992. [CrossRef]
15. Zhu, Z.Q.; Chen, Z.T.; Zhang, Y. A novel polishing technology for leading and trailing edges of aero-engine blade. *Int. J. Adv. Manuf. Technol.* **2021**, *116*, 1871–1880. [CrossRef]
16. Fu, Y.Z.; Gao, H.; Yan, Q.S.; Wang, X.P. A new predictive method of the finished surface profile in abrasive flow machining process. *Precis. Eng.-J. Int. Soc. Precis. Eng. Nanotechnol.* **2019**, *60*, 497–505. [CrossRef]
17. Wu, B.; Yi, R.; Ding, X.M.; Chiu, T.; He, Q.P.; Deng, H. Surface evolution mechanism for atomic-scale smoothing of Si via atmospheric pressure plasma etching. *J. Manuf. Process.* **2024**, *132*, 353–362. [CrossRef]
18. Zhou, Y.Q.; Xu, K.Z.; Gao, Y.H.; Yu, Z.N.; Zhu, F.L. Effects of oxidizer concentration and abrasive type on interfacial bonding and material removal in 4H-SiC polishing processes. *Phys. Chem. Chem. Phys.* **2024**, *26*, 27791–27806. [CrossRef]
19. Zhai, Q.; Zhai, W.J.; Deng, T.H. Study on process optimization of ultrasound assisted magneto-rheological polishing of sapphire hemisphere surface based on Fe_3O_4/SiO_2 core-shell abrasives. *Tribol. Int.* **2023**, *181*, 20. [CrossRef]
20. Ye, L.; Wu, J.; Zhu, X.; Liu, Y.; Li, W.; Chuai, S.; Wang, Z. Optimization of polishing fluid composition for single crystal silicon carbide by ultrasonic assisted chemical-mechanical polishing. *Sci. Rep.* **2024**, *14*, 13. [CrossRef]
21. Cheng, J.; Lv, Y.R.; Zhang, F.; Han, P.; Miao, Q.H.; Huang, Z.X. Understanding the adsorption mechanism of benzotriazole and its derivatives as effective corrosion inhibitors for cobalt in chemical mechanical polishing. *Appl. Surf. Sci.* **2025**, *682*, 11. [CrossRef]
22. Lv, G.; Liu, S.S.; Cao, Y.X.; Zhang, Z.F.; Li, X.F.; Zhang, Y.F.; Liu, T.; Liu, B.S.; Wang, K.Y. Effect of synergistic CeO2/MoS2 abrasives on surface roughness and material removal rate of quartz glass. *Ceram. Int.* **2024**, *50*, 48671–48679. [CrossRef]
23. Peng, C.; Gao, H.; Wang, X.P. On Characterization of Shear Viscosity and Wall Slip for Concentrated Suspension Flows in Abrasive Flow Machining. *Materials* **2023**, *16*, 6803. [CrossRef] [PubMed]
24. Kum, C.W.; Wu, C.H.; Wan, S.; Kang, C.W. Prediction and compensation of material removal for abrasive flow machining of additively manufactured metal components. *J. Mater. Process Technol.* **2020**, *282*, 13. [CrossRef]
25. Fu, Y.Z.; Wang, X.P.; Gao, H.; Wei, H.B.; Li, S.C. Blade surface uniformity of blisk finished by abrasive flow machining. *Int. J. Adv. Manuf. Technol.* **2016**, *84*, 1725–1735. [CrossRef]
26. Guo, L.; Wang, X.; Lyu, B.; Cao, L.; Mao, Y.; Wang, J.; Chen, H.; Wang, J.; Yuan, J. Shear-thickening polishing of inner raceway surface of bearing and suppression of edge effect. *Int. J. Adv. Manuf. Technol.* **2022**, *121*, 4055–4068. [CrossRef]
27. Lu, M.M.; Yang, Y.K.; Lin, J.Q.; Du, Y.S.; Zhou, X.Q. Research progress of magnetorheological polishing technology: A review. *Adv. Manuf.* **2024**, *12*, 642–678. [CrossRef]
28. Wang, Y.Q.; Yin, S.H.; Hu, T. Ultra-precision finishing of optical mold by magnetorheological polishing using a cylindrical permanent magnet. *Int. J. Adv. Manuf. Technol.* **2018**, *97*, 3583–3594. [CrossRef]
29. Luo, Z.; Zhang, Z.; Zhao, F.; Fan, C.; Feng, J.; Zhou, H.; Meng, F.; Zhuang, X.; Wang, J. Advanced polishing methods for atomic-scale surfaces: A review. *Mater. Today Sustain.* **2024**, *27*, 29. [CrossRef]
30. Zhu, W.L.; Beaucamp, A. Generic three-dimensional model of freeform surface polishing with non-Newtonian fluids. *Int. J. Mach. Tools Manuf.* **2022**, *172*, 20. [CrossRef]
31. Zhu, W.L.; Beaucamp, A. Non-Newtonian fluid based contactless sub-aperture polishing. *CIRP Ann.-Manuf. Technol.* **2020**, *69*, 293–296. [CrossRef]
32. Singh, P.; Singh, L.; Singh, S. A review on magnetically assisted abrasive flow machining and abrasive material type. *Proc. Inst. Mech. Eng. Part E-J. Process Mech. Eng.* **2022**, *236*, 2765–2781. [CrossRef]
33. Rajput, A.S.; Das, M.; Kapil, S. A comprehensive review of magnetorheological fluid assisted finishing processes. *Mach. Sci. Technol.* **2022**, *26*, 339–376. [CrossRef]
34. Liu, H.; Wang, J.; Huang, C.Z. Abrasive liquid jet as a flexible polishing tool. *Int. J. Mater. Prod. Technol.* **2008**, *31*, 2–13. [CrossRef]

35. Gupta, K.; Gupta, M.K. Developments in nonconventional machining for sustainable production: A state-of-the-art review. *Proc. Inst. Mech. Eng. Part C-J. Eng. Mech. Eng. Sci.* **2019**, *233*, 4213–4232. [CrossRef]
36. Wen, S.Z.; Huang, P.; Tian, Y.; Ma, L.R. *Principles of Tribology*; Tsinghua University: Beijing, China, 2018.
37. Guo, L.; Wang, X.; Lyu, B.; Lyu, J.; Wang, J.; Chen, H.; Zhao, W.; Yuan, J. Ultra-precision machining process of inner surface considering shear-thickening polishing method. *Proc. Inst. Mech. Eng. Part B-J. Eng. Manuf.* **2024**, *239*, 264–277. [CrossRef]
38. Guo, L.G.; Dai, Z.H.; Wang, D.F.; Wang, X.; Lyu, B.H.; Yuan, J.L. Effects of Force Rheological Polishing Processes on Surface Quality and Accuracy of Bearing Raceway. *China Mech. Eng.* **2025**, *36*, 271–279.
39. Shakiba, M.; Ghomi, E.R.; Khosravi, F.; Jouybar, S.; Bigham, A.; Zare, M.; Abdouss, M.; Moaref, R.; Ramakrishna, S. Nylon-A material introduction and overview for biomedical applications. *Polym. Adv. Technol.* **2021**, *32*, 3368–3383. [CrossRef]
40. Yang, Q.Q.; Huang, P.; Fang, Y.F. A novel Reynolds equation of non-Newtonian fluid for lubrication simulation. *Tribol. Int.* **2016**, *94*, 458–463. [CrossRef]
41. Yang, Q.Q.; Chen, Y.J.; Huang, P. A Novel Method to Determine EHL Film Thickness with Optical Interference. *Appl. Mech. Mater.* **2013**, *456*, 549–554. [CrossRef]
42. Huang, P.; Yang, Q.Q. Theory and contents of frictional mechanics. *Friction* **2014**, *2*, 27–39. [CrossRef]
43. Wei, H.B.; Gao, H.; Wang, X.Y. Development of novel guar gum hydrogel based media for abrasive flow machining: Shear-thickening behavior and finishing performance. *Int. J. Mech. Sci.* **2019**, *157*, 758–772. [CrossRef]

Disclaimer/Publisher's Note: The statements, opinions and data contained in all publications are solely those of the individual author(s) and contributor(s) and not of MDPI and/or the editor(s). MDPI and/or the editor(s) disclaim responsibility for any injury to people or property resulting from any ideas, methods, instructions or products referred to in the content.

micromachines

MDPI

Article

Research on Helical Electrode Electrochemical Drilling Assisted by Anode Vibration for Jet Micro-Hole Arrays on Tube Walls

Tao Yang [1,2,3,*], Yikai Xiao [1], Yusen Hang [1,2], Xiujuan Wu [1,3] and Weijing Kong [1,2]

1 College of Mechanical Engineering, Nanjing Vocational University of Industry Technology, Nanjing 210023, China; xyk13770725152@163.com (Y.X.); 2023101416@niit.edu.cn (Y.H.); wuxj@niit.edu.cn (X.W.); kongwj@nuaa.edu.cn (W.K.)
2 Jiangsu Key Laboratory of Precision and Micro-Manufacturing Technology, Nanjing University of Aeronautics and Astronautics, Nanjing 210016, China
3 Precision Manufacturing Engineering and Technology Research Center of Jiangsu Province, Nanjing Vocational University of Industry Technology, Nanjing 210023, China
* Correspondence: yangtao860632@163.com

Abstract: The electrochemical cutting technique, utilizing electrolyte flushing through micro-hole arrays in the radial direction of a tube electrode, offers the potential for cost-effective and high-surface-integrity machining of large-thickness, straight-surface structures of difficult-to-cut materials. However, fabricating the array of jet micro-holes on the tube electrode sidewall remains a significant challenge, limiting the broader application of this technology. To enhance the efficiency and quality of machining these jet micro-holes on the tube sidewall, a helical electrode electrochemical drilling method assisted by anode vibration has been proposed. The influence of parameters, such as the rotational direction and speed of the helical electrode, as well as the vibration amplitude and frequency of the workpiece, on the machining results was investigated using fluid field simulation and machining experiments. It was found that these auxiliary movements could facilitate the renewal of electrolytes within the machining gap, thereby enhancing the efficiency and quality of electrochemical drilling. Using the optimized machining parameters, an array of 10 jet micro-holes with a diameter of 200 μm was machined on the metal tube sidewall. Electrochemical cutting with radial electrolyte flushing tests were then performed through these micro-holes.

Keywords: jet micro-hole arrays; electrochemical drilling; helical electrode; anode vibration

1. Introduction

Electrochemical cutting is a machining process based on the principle of electrochemical anodic dissolution, wherein a wire or rod tool electrode dissolves and removes localized material from the workpiece to complete the cut [1]. As a form of electrochemical machining, it offers high precision, excellent surface quality, tool electrode wear resistance, and no machining stress [2]. It is widely used for cutting and fabricating small precision components [3]. Furthermore, research is actively exploring its use for machining large-thickness ruled surface components, such as aviation turbine disk mortise structures, while maintaining high surface integrity [4,5].

A key limiting factor in electrochemical cutting is the restricted mass transport within the machining gap, which hampers both machining efficiency and quality [6]. To overcome this, a radial electrolyte flushing electrochemical cutting method using a tube electrode with an array of holes has been proposed. This method replaces traditional wire or rod electrodes with a metal tube featuring an array of jet holes in its sidewall. The electrolyte is injected rapidly through these arrayed jet holes into the machining gap, significantly improving mass transport and enhancing cutting efficiency [7]. The structure of the arrayed jet micro-holes on the tube sidewall plays a crucial role in the overall machining process [8], making their precise fabrication essential for the success of this technique [9].

Several methods exist for machining micro-holes, including micro-drilling, laser drilling, electrical discharge drilling, and electrochemical drilling. Micro-drilling removes material through the shear force generated by a micro drill, offering high machining efficiency [10]. However, drilling arrayed jet micro-holes on a tube wall is challenging due to tool inclination and potential breakage, especially in the early stages. Additionally, chips may enter the metal cavity, and burrs can form at the edges of the holes, interfering with subsequent electrolyte injection [11]. Laser drilling, which removes material through the localized melting and vaporization induced by high-energy laser beams, avoids these issues, as it does not require traditional drilling tools, thus eliminating tool wear. This non-contact process also avoids machining stresses, achieving high precision and efficiency for a variety of materials [12]. Similarly, electrical discharge drilling uses spark thermal energy to remove material, offering high efficiency and no contact between the tool and workpiece. It can also break through the material's strength and hardness limits [13]. However, both laser and electrical discharge drilling introduce thermal defects, such as deformation, heat-affected zones, and recast layers, which may compromise the integrity of the drilled micro-holes and affect the tube's subsequent performance [14].

Electrochemical drilling is a machining technique that removes material from a workpiece via electrochemical anode dissolution. In this process, localized material is dissolved in ionic form, ensuring high machining precision and a surface free from defects like burrs, recast layers, and heat-affected zones. Since the tool electrode does not physically contact the workpiece, processing stress is eliminated. Only hydrogen evolution occurs at the tool electrode surface, preventing tool wear [15]. This method is particularly advantageous for the high-quality manufacturing of jet micro-holes in tube walls [16]. For instance, micro-holes with a diameter of approximately 400 μm have been successfully machined in a 4 mm thick superalloy plate [17]. The use of a helical drill as the tool electrode significantly enhances both machining efficiency and micro-hole precision [18,19]. Liu et al. achieved a non-tapered micro-hole with a diameter of 186 μm in a 500 μm thick GH4169 plate by using a rotating helical electrode and ultra-short voltage pulses [20]. However, the removal efficiency of electrolytic products within the end-face machining gap between the end face of the helical electrode and the bottom of the hole still needs to be improved. To address this, Wang et al. utilized ultrasound to cause high-frequency vibration of the tool electrode, thereby facilitating the removal of products within the end-face machining gap. Ultimately, they were able to machine a deep hole with a depth of 5.4 mm and a depth-to-diameter ratio of 12.3 [21]. In this method, an excessively high ultrasonic vibration frequency can easily impact machining accuracy.

To improve the efficiency and quality of jet micro-hole processing on tube walls, we propose a helical electrode electrochemical drilling method assisted by anode vibration. The electrolyte within the machining gap is agitated by the high-speed rotational movement of the helical electrode, while the electrolyte in the end-face machining gap is intermittently squeezed by the vertical vibration of the anode. This promotes the flow of the electrolyte, accelerates the removal of electrolytic products and the renewal of the electrolyte, and consequently enhances machining efficiency and quality. Simulations and experiments were conducted to investigate the effects of the helical electrode's rotational movement and the anode's vibrational movement, and optimal machining parameters were selected. An array of jet micro-holes was machined on the sidewall of a metal tube, and a radial electrolyte flushing electrochemical cutting experiment was conducted using this tube with jet micro-holes as the electrode.

2. Machining Principle and Simulation Analysis

2.1. Machining Principle

Figure 1 illustrates the principle of helical electrode electrochemical drilling assisted by anode vibration on a metal tube sidewall. A tungsten steel alloy twist micro-drill is chosen as the helical electrode, and a stainless-steel 304 tube serves as the workpiece immersed in the electrolyte. The helical electrode is connected to the negative terminal, while the metal tube is connected to the positive terminal of the pulse power supply. Under the applied pulse voltage, an electrochemical reaction occurs between the tube wall and the helical electrode, resulting in the dissolution of localized material from the tube. As the helical electrode feeds, a micro-hole structure is formed on the tube sidewall. The high-speed rotation of the electrode stirs the electrolyte, promoting its renewal within the machining gap. The piezoelectric ceramic unit drives the tube to vibrate vertically, further enhancing the electrolyte flow and improving both machining stability and efficiency.

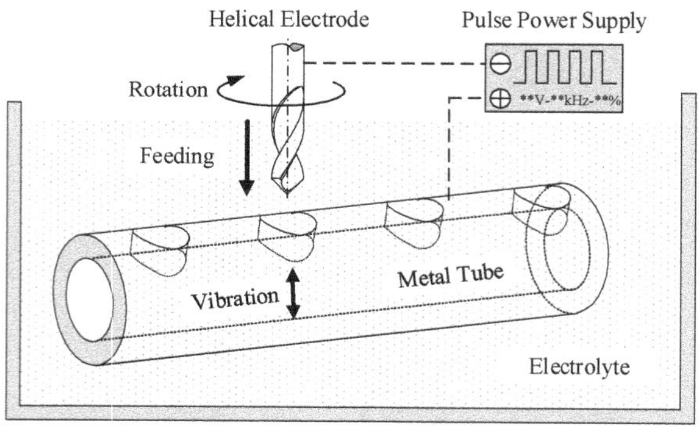

Figure 1. Principle schematic diagram of helical electrode electrochemical drilling assisted by anode vibration. ** is the specific electrical parameter value during processing.

2.2. Simulation Analysis

A simulation study was conducted to examine the impact of the helical electrode's rotational motion and the workpiece's vertical vibration on the machining process, particularly the electrolyte flow field within the machining gap. Figure 2 shows the simulation model, with parameters detailed in Table 1. To clarify the electrolyte flow dynamics, a vertical section (A) was chosen as a reference.

Table 1. Simulation parameters.

Parameter	Value
Diameter of helical electrode (mm)	0.1
Rotational speed (rpm)	0, 3000, −3000
Vibration amplitude (μm)	10
Vibration frequency (Hz)	100
Diameter of hole (mm)	0.2
Depth of hole (mm)	0.1
End-face machining gap (μm)	20

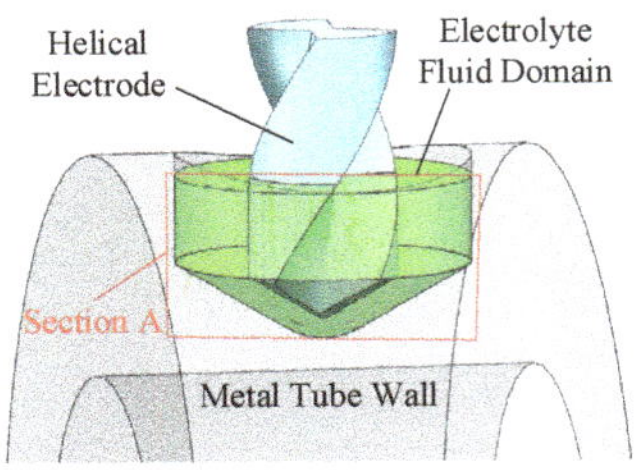

Figure 2. Simulation model of the electrolyte flow field.

During the simulation process, the rotational speed of the helical electrode was set to 0 and 3000 rpm, respectively, to compare and analyze the impact of the helical electrode's rotation on the electrolyte flow field. Additionally, the rotational speed of the helical electrode was set to −3000 rpm and 3000 rpm to compare and analyze the influence of the helical electrode's rotational direction on the electrolyte flow field. Figure 3 presents the velocity contour of the electrolyte flow during the high-speed rotation of the helical electrode. Without rotation, the electrolyte remains stationary in the machining gap. However, at 3000 rpm, the helical electrode stirs the electrolyte, creating a localized flow that aids in the removal of electrolytic products and the renewal of the electrolyte. Furthermore, the velocity distribution of the electrolyte remains the same regardless of the direction of rotation. However, the electrolyte remains stagnant between the bottom surface of the tool and the bottom surface of the hole (the end-face machining gap). To address this, the application of workpiece vibration is proposed, aiming to drive the electrolyte flow by altering the size of the machining gap.

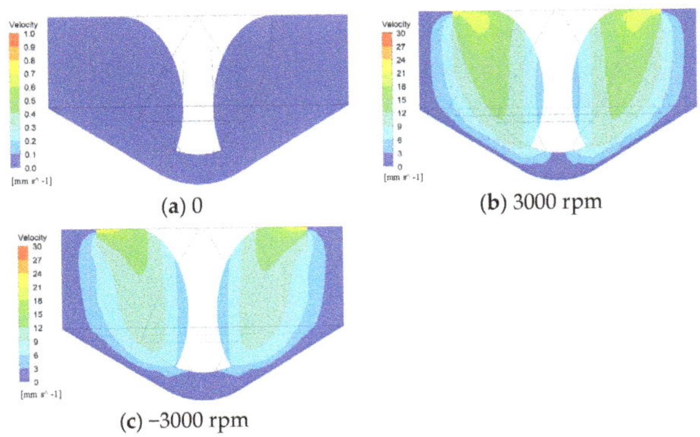

Figure 3. Distribution contour of electrolyte flow velocity within the machining gap during rotation of the helical electrode.

During the up and down vibration of the workpiece, when the workpiece moves upwards, the end-face machining gap narrows, exerting a compressive force on the electrolyte. Conversely, when the workpiece moves downward, the end-face machining gap widens, creating a suction effect on the electrolyte. Therefore, the simulation focused on exploring both the upward and downward vibration processes of the workpiece. The amplitude and frequency of the vibration were set to 10 μm and 100 Hz, respectively. Figure 4 demonstrates the effect of workpiece vibration on the electrolyte flow field. Vertical movement of the workpiece causes periodic expansion and contraction of the machining gap, generating a pumping and squeezing effect that drives the electrolyte flow. This accelerates the removal of electrolytic products and promotes electrolyte renewal.

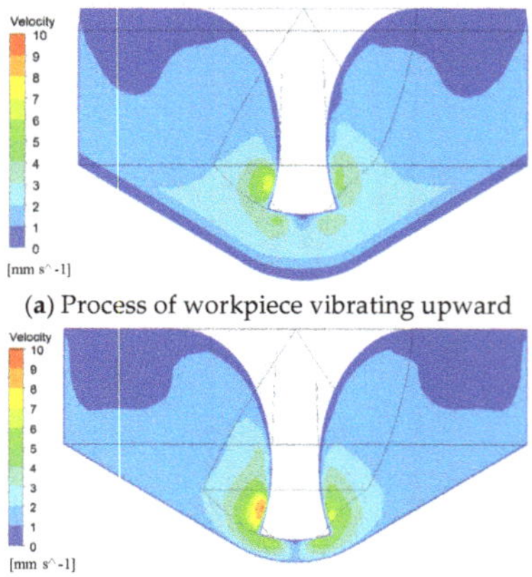

(**a**) Process of workpiece vibrating upward

(**b**) Process of workpiece vibrating downwards

Figure 4. Distribution contour of electrolyte flow velocity within the machining gap during workpiece vibration.

3. Experimental Details

Figure 5 shows the experimental setup for helical electrode electrochemical drilling assisted by anode vibration. The system includes a motion control system, an X/Y/Z precision motion platform (M511.DG, PI, Karlsruhe, Germany), an electric spindle (NR-2551, NSK, Tokyo, Japan), a piezoelectric ceramic unit (P158.ZCD, PI, Karlsruhe, Germany), a pulse power supply (YS9000D, Yisheng, Shanghai, China), an oscilloscope (DS6104, RIGOL, Suzhou, China), a Charge-Coupled Device (CCD) vision system, and a dedicated fixture and electrolyte tank. The stainless-steel tube is secured in the fixture within the electrolyte tank, where it is immersed in the electrolyte. The helical electrode, mounted on the electric spindle, undergoes high-speed rotation. The X/Y/Z precision motion platform controls the electrode's feed motion, while the piezoelectric ceramic units drive the tube's vertical vibration. The pulse power supply operates at a frequency of 1–100 kHz, ensuring both precision and efficiency. The positive and negative terminals of the power supply are connected to the tube and the helical electrode, respectively. The oscilloscope monitors the voltage and current in the circuit while the CCD vision system observes and records the machining process. Specific experimental parameters are listed in Table 2.

Table 2. Experimental parameters.

Parameter	Value
Stainless-steel tube	Outer diameter 0.7 mm Inner diameter 0.4 mm
Helical electrode	Diameter 0.1 mm Screw pitch 0.48 mm
Electrolyte	$NaNO_3$ solution 50 g/L
Electrical parameter	8 V-25%-100 kHz
Feed rate (μm/s)	0.8
Feed quantity (μm)	200
Helical electrode rotation speed (rpm)	1000, 2000, 3000, 4000, 5000
Workpiece vibration amplitude (μm)	0, 2, 4, 6, 8, 10
Workpiece vibration frequency (Hz)	50, 75, 100, 125, 150

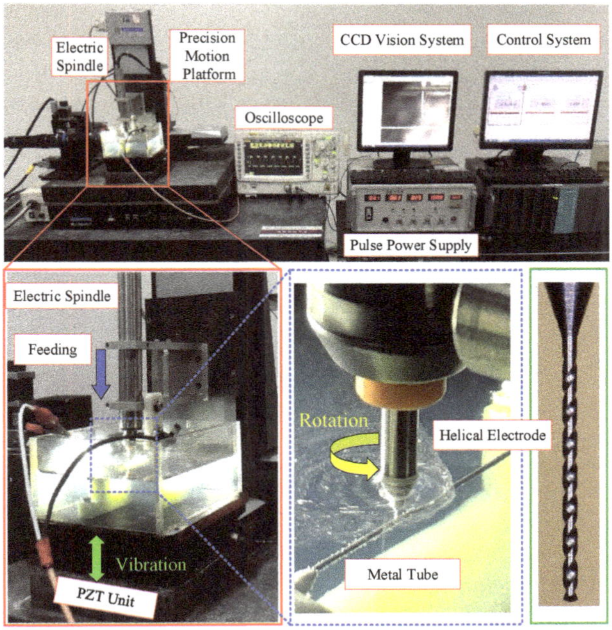

Figure 5. Experimental setup of helical electrode electrochemical drilling assisted by anode vibration.

During electrochemical drilling, the helical electrode's rotational motion and the workpiece's vertical vibration influence electrolyte flow within the machining gap, affecting bubble removal and the discharge of insoluble electrolytic products. As outlined in Equation (1), variations in product removal efficiency alter the volume fraction of electrolytic products in the electrolyte, which in turn impacts its conductivity. According to Equation (2), when the helical electrode's feed rate is constant, changes in electrolyte conductivity affect the end-face machining gap. In a balanced machining state, the relationship between the side-face and end-face machining gaps is defined by Equation (3). Equation (4) correlates the drilled hole size with the side-face machining gap. In this experiment, a constant feed rate is used, and the effects of the helical electrode's rotational movement and the workpiece's vertical vibration on machining results are analyzed by comparing hole diameters produced under varying auxiliary motion parameters.

$$\kappa = \kappa_0 \frac{2(1-\beta)}{2+\beta} \tag{1}$$

$$\Delta_b = \frac{\eta \omega \kappa U_R}{v_b} \tag{2}$$

$$\Delta_s \cong \Delta_b \sqrt{1 + \frac{2D_{TE}}{\Delta_b}} \tag{3}$$

$$D_H = D_{TE} + 2\Delta_s \tag{4}$$

Among them, κ is the conductivity of the electrolyte, κ_0 is the initial conductivity of the electrolyte, β is the content of electrolytic products in the electrolyte, Δ_b is the end-face machining gap, η is the current efficiency, ω is the volumetric electrochemical equivalent,

U_R is the voltage between the two electrodes, v_b is the feed rate of the tool electrode, Δ_s is the side machining gap, D_{TE} is the diameter of the tool electrode, and D_H is the diameter of the hole.

Ten holes are drilled for each set of machining parameters. After drilling, the stainless-steel tube is cleaned with ultrasonic waves in alcohol. The morphology of the micro-holes is captured using a digital microscope (DVM5000, Leica, Wetzlar, Germany), and the diameters are measured with an optical microscope (SMT7-SFA, Olympus, Tokyo, Japan). Hole diameters are measured for each hole to calculate the average diameter and standard deviation.

4. Results and Discussion

4.1. *Influence of the Helical Electrode Rotation*

This study investigates the influence of the rotation speed and direction of the helical electrode on the diameter of drilled jet micro-holes, with the workpiece vibration amplitude set at 8 μm and frequency at 100 Hz. The results are presented in Figure 6. Regardless of rotation direction, an increase in the rotation speed of the helical electrode leads to a gradual enlargement of the hole diameter. As indicated by the flow field simulation results (Figure 3), when the helical electrode rotates, the helical groove structure on its surface generates a stirring effect on the electrolyte within the machining gap, enhancing the flow of the electrolyte. This facilitates the rapid removal of bubbles and insoluble electrolytic products generated by the electrochemical reaction from the machining area. Additionally, as the rotational speed increases, the stirring effect of the helical electrode on the electrolyte becomes more significant. The removal rate of bubbles and insoluble electrolytic products from the micro-hole accelerates. Based on Equations (1)–(4), it can be inferred that the conductivity of the electrolyte in the machining gap is relatively high, and the removal amount of sidewall materials increases, which makes the side machining gap larger, and thus the micro-hole diameter becomes larger.

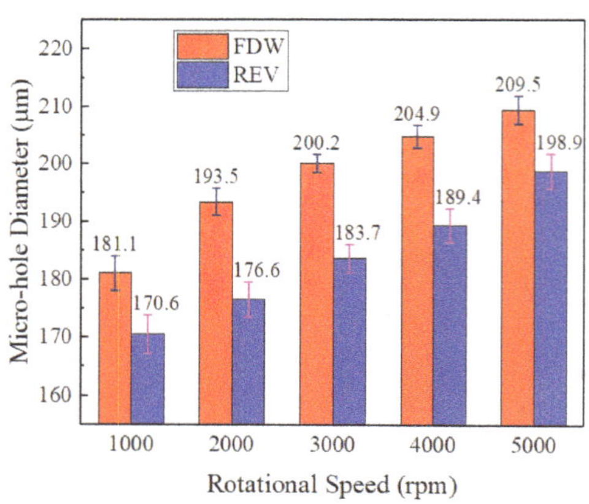

Figure 6. Diameter of jet holes processed by helical electrode under different rotation states.

For a given rotation speed, the hole diameter is larger when the helical electrode rotates in the forward direction compared to when it rotates in reverse. This is because, during the actual machining process, bubbles and insoluble electrolytic products are generated. These exist in the electrolyte and affect its conductivity, particularly the bubbles. The bubbles are produced on the surface of the helical electrode. In reverse rotation, the axial force exerted on the electrolyte is downward, with fresh electrolyte entering the hole along the electrode wall and exiting the machining gap after reaching the hole's bottom (Figure 7a). In this flow pattern, hydrogen bubbles from the electrode wall are carried into the flow by the fresh electrolyte, reducing the electrolyte's conductivity and resulting in a smaller hole diameter. In contrast, when rotating forward, the axial force is upward, causing spent electrolytes to be extracted along the electrode wall while fresh electrolyte enters the hole, maintaining the electrolytic reaction at the bottom of the hole (Figure 7b). This flow pattern facilitates the removal of hydrogen bubbles, wherein the bubbles precipitated from the wall of the helical electrode are directly carried out of the small hole by the electrolyte along its wall, resulting in fewer hydrogen bubbles in the electrolyte throughout the entire flow process, improving electrolyte conductivity and leading to a larger hole diameter.

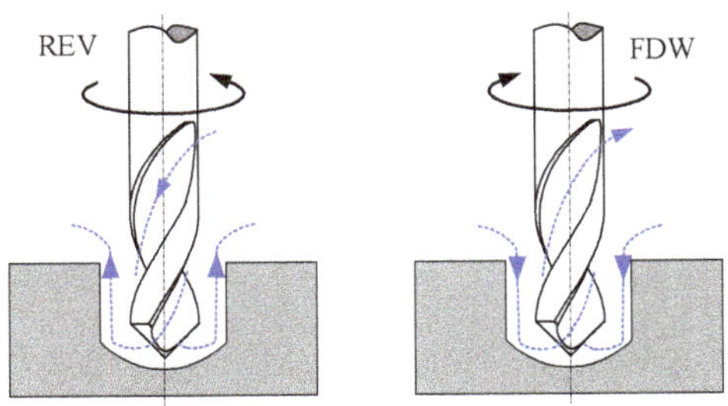

Figure 7. Schematic diagram of electrolyte renewal paths within the machining gap for different rotational directions of the electrode. (**a**) helical electrode rotating reversal, (**b**) helical electrode rotating forward.

4.2. Influence of Workpiece Vibration Amplitude

With the helical electrode rotating forward at 3000 rpm and the workpiece vibrating at 100 Hz, Figure 8 shows hole diameters for varying vibration amplitudes. At zero vibration (amplitude = 0), the smallest hole diameter and the highest standard deviation are observed. This is because, during the electrochemical drilling process, the feed rate of the helical electrode is the same as the material removal rate at the bottom of the micro-hole. As a result, the end-face machining gap between the end face of the helical electrode and the bottom of the micro-hole remains constant. The electrolyte remains relatively stationary, causing slower removal of internal bubbles and insoluble electrolytic products. This leads to a relatively low electrolyte conductivity. According to Equations (2)–(4), a smaller side machining gap results in a smaller diameter of the drilled hole. Meanwhile, the accumulation of a large number of electrolytic products within the end-face machining gap can easily cause variations in the electrolyte conductivity, leading to reduced aperture consistency and larger deviation values of micro-hole diameter.

When the workpiece vibrates upward and downward, the end-face machining gap between the end face of the helical electrode and the bottom of the micro-hole changes periodically. Based on the simulation results (Figure 4), it can be observed that when the workpiece vibrates upward, the end-face machining gap decreases, exerting a compressive effect on the electrolyte. Conversely, when the workpiece vibrates downward, the end-face machining gap increases, creating a suction effect on the electrolyte. The alternating occurrence of these two effects generates localized flow in the electrolyte, which facilitates the removal of electrolytic products within the machining gap. This results in a relatively high electrolyte conductivity and larger micro-hole diameters. Furthermore, due to the timely renewal of the electrolyte, the conductivity remains relatively constant, leading to better consistency in the micro-hole diameters produced.

As vibration amplitude increases, the hole diameter increases. This is because when the vibration frequency remains constant, an increase in vibration amplitude accelerates the movement speed of the workpiece, intensifying the compression and suction effects on the electrolyte within the end-face machining gap. This enhances the removal efficiency of electrolytic products. Consequently, the conductivity of the electrolyte increases relatively, leading to an increase in the amount of electrolytic removal and resulting in larger micro-hole diameters. However, beyond an amplitude of 10 μm, the hole diameter decreases. This is due to excessive vibration reducing the time the electrode is in close contact with the hole bottom, limiting the electrolytic reaction duration and thus reducing material removal and hole size.

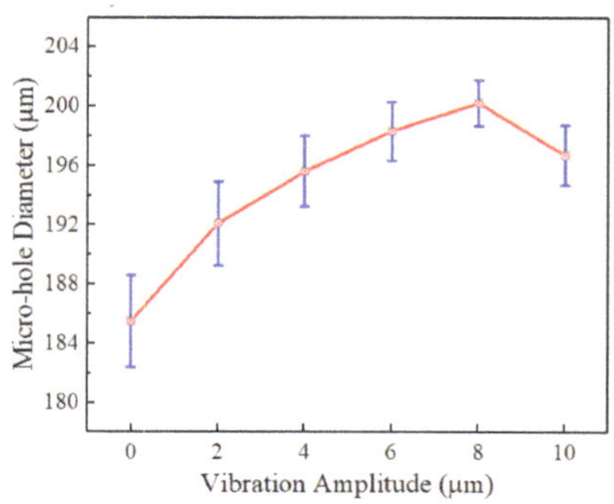

Figure 8. The diameter of the jet hole processed under different vibration amplitudes.

4.3. Influence of Workpiece Vibration Frequency

When the helical electrode rotates forward at 3000 rpm, the workpiece vibrates with an amplitude of 8 μm, and Figure 9 presents hole diameters at varying vibration frequencies during the workpiece's vibration. As frequency increases from 50 Hz to 100 Hz, the hole diameter increases from 190 μm to 200 μm. This is because when the vibration amplitude remains constant, an increase in vibration frequency leads to an increase in the number of vibrations of the workpiece per unit time, which accelerates its movement speed. This, in turn, enhances the compression and suction effects on the electrolyte within the end-face machining gap. As a result, the renewal efficiency of the electrolyte in the machining gap is improved, allowing electrolytic products to be quickly removed. This increases the conductivity of the electrolyte and results in larger micro-hole diameters being produced.

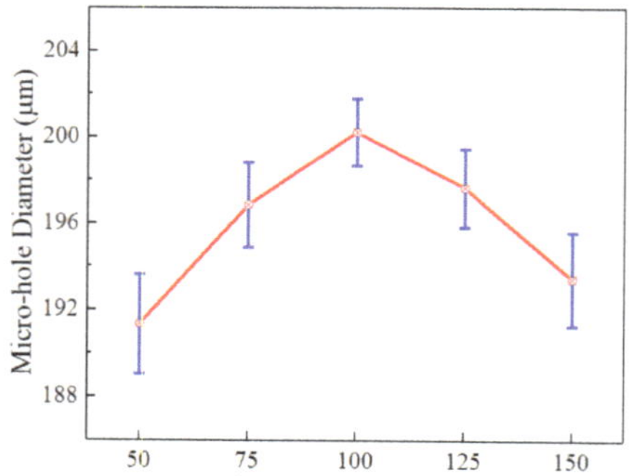

Figure 9. The diameter of the jet hole processed under different vibration frequencies.

However, when the vibration frequency increases from 100 Hz to 150 Hz, the machined jet micro-hole diameter decreases from 200 μm to 185 μm. This is because the number of vibrations of the workpiece per unit of time increases when the vibration frequency increases, and the vibration time of a single time is shortened, causing the workpiece to vibrate too quickly. This results in insufficient time for the diffusion of the electrolytic products in the machining gap, hindering the renewal of the electrolyte. Therefore, the machined jet hole diameter decreases, and the consistency of the hole diameter is poor.

4.4. Fabrication of an Array of Jet Micro-Holes on the Wall of a Stainless-Steel Tube

Based on the research outlined above, the helical electrode was set to rotate forward at 3000 rpm. The workpiece vibration amplitude and frequency were set to 8 μm and 100 Hz, respectively. Other processing parameters are listed in Table 2. A total of 10 arrayed jet holes, spaced 0.5 mm apart, were machined into the side wall of the stainless-steel tube, as shown in Figure 10. The average hole diameter was 200 μm, with a standard deviation of 2.97 μm.

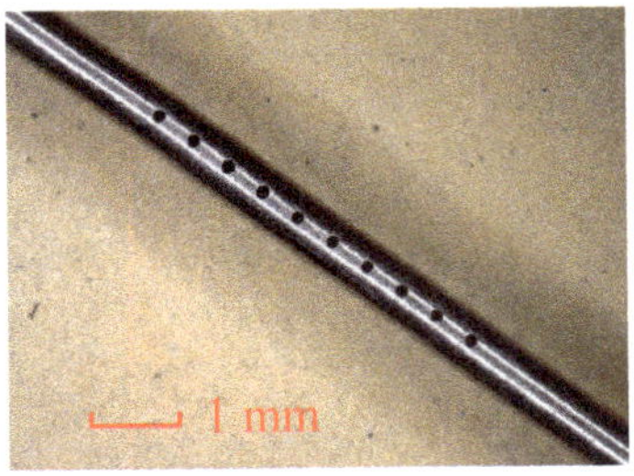

Figure 10. A jet micro-hole array prepared on the wall of a stainless-steel tube.

4.5. Application of Tube Electrode with a Jet Micro-Hole Array in Electrochemical Cutting

This metal tube, featuring arrayed jet micro-holes on its sidewall, was used as the tool electrode for radial electrolyte flushing electrochemical cutting experiments on a 5 mm thick stainless-steel 304 plate. The machining principle is illustrated in Figure 11, and the spraying behavior of the electrolyte from the array of jet micro-holes is depicted in Figure 12. To enhance the uniformity and consistency of the electrolyte flow within the machining gap, the workpiece underwent reciprocal motion along the tube electrode's axial direction. The amplitude of this motion was 0.5 mm with a frequency of 1.5 Hz. Additional machining parameters are provided in Table 3. A slit structure with an average width of 1.089 mm and a standard deviation of 35.3 μm was successfully machined, as shown in Figure 13.

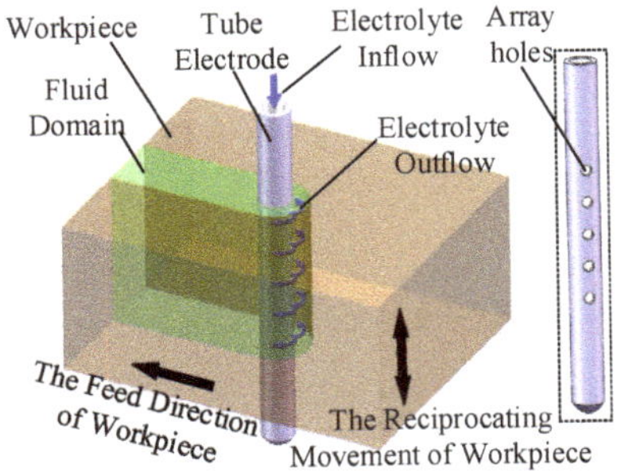

Figure 11. Schematic diagram of electrochemical cutting with radial electrolyte flushing [7].

Figure 12. The spraying state of the electrolyte from the array of jet micro-holes in the tube wall.

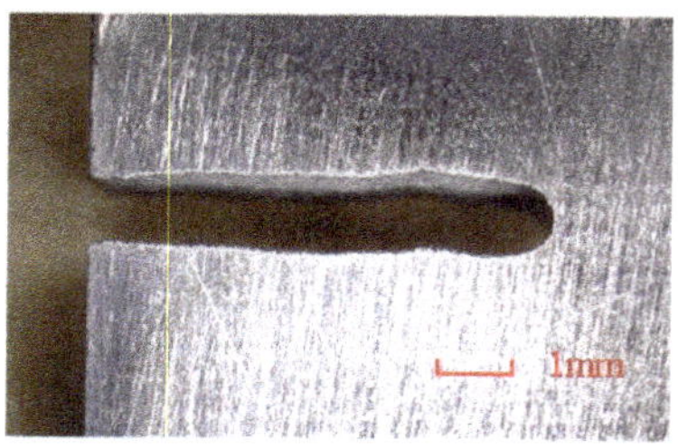

Figure 13. The slit structure produced by electrochemical cutting using this metal tube with jet micro-hole arrays on tube walls.

Table 3. Machining parameters of electrochemical cutting with radial electrolyte flushing.

Parameter	Value
Electrical parameter	12 V-35%-50 kHz
Electrolyte type	$NaNO_3$ solution
Electrolyte concentration (g/L)	100
Inlet pressure (MPa)	2.0
Feed rate (μm/s)	4

5. Conclusions

To achieve efficient and high-quality processing of jet micro-holes on the sidewall of metal tubes, a method combining helical electrode electrochemical drilling with anode vibration has been proposed. By simulating the electrolyte flow field within the machining gap and conducting machining experiments, the effects of the helical electrode's rotation direction and speed, as well as the vibration amplitude and frequency of the workpiece, on the jet micro-hole diameter were analyzed. The following conclusions were drawn:

(1) High-speed rotation of the helical electrode and vibration of the workpiece enhance electrolyte flow within the machining gap, promoting rapid removal of electrolytic products and efficient renewal of the electrolyte;

(2) Forward rotation of the helical electrode results in larger jet micro-hole diameters compared to reverse rotation. Moreover, as the rotation speed increases, the diameter of the machined jet micro-holes also increases;

(3) Workpiece vibration leads to larger jet micro-hole diameters compared to non-vibrated processing. At optimal vibration amplitudes and frequencies, the jet micro-holes exhibit larger diameters and smaller deviations;

(4) Using the optimal machining parameters, 10 jet micro-holes were successfully machined on the stainless-steel tube wall. A radial electrolyte flushing electrochemical

cutting experiment was conducted using this tube with jet micro-holes as the electrode, resulting in the formation of a slit structure on a 5 mm thick stainless-steel 304 plate.

Author Contributions: Conceptualization, T.Y.; methodology, T.Y.; software, Y.X.; validation, Y.H.; formal analysis, Y.X.; investigation, X.W.; resources, T.Y.; data curation, W.K.; writing—original draft preparation, T.Y.; writing—review and editing, T.Y.; visualization, Y.H.; supervision, Y.H.; project administration, T.Y.; funding acquisition, X.W. All authors have read and agreed to the published version of the manuscript.

Funding: This research was supported by the National Natural Science Foundation of China (No. 52205473), the Natural Science Foundation of Jiangsu Province (No. BK20220688), the China Postdoctoral Science Foundation (No. 2023M741683), the Open Fund for Jiangsu Key Laboratory of Precision and Micro-Manufacturing Technology (No. JSKL2122K02), the Foundation of the Jiangsu Province Precision Manufacturing Engineering and Technology Research Center (No. 05SJGCZX-01), the Start-up Fund for New Talented Researchers of Nanjing Vocational University of Industry Technology (No. YK22-01-01), and Jiangsu University's "Blue Project".

Data Availability Statement: The original contributions presented in this study are included in the article. Further inquiries can be directed to the corresponding author.

Conflicts of Interest: The authors declare no conflicts of interest.

References

1. Zhu, D.; Wang, K.; Qu, N.S. Micro Wire electrochemical cutting by using in situ fabricated Wire electrode. *CIRP Ann. Manuf. Technol.* **2007**, *56*, 241–244. [CrossRef]
2. Sharma, V.; Patel, D.S.; Jain, V.K.; Ramkumar, J. Wire electrochemical micromachining: An overview. *Int. J. Mach. Tool. Manuf.* **2020**, *155*, 103579. [CrossRef]
3. Bi, X.L.; Zeng, Y.B.; Qu, N.S. Wire electrochemical micromachining of high-quality pure-nickel microstructures focusing on different machining indicators. *Precis. Eng.* **2020**, *16*, 14–22. [CrossRef]
4. Klocke, F.; Herrig, T.; Zeis, M.; Klink, A. Experimental investigations of cutting rates and surface integrity in wire electrochemical machining with rotating electrode. *Proc. CIRP* **2018**, *68*, 725–730. [CrossRef]
5. Fang, X.L.; Han, Z.; Chen, M.; Zhu, D. Pulse-current wire electrochemical machining with axial electrolyte flushing along a rotating helical Wire tool. *J. Electrochem. Soc.* **2020**, *167*, 113503. [CrossRef]
6. Debnath, S.; Doloi, B.; Bhattacharyya, B. Review—Wire Electrochemical Machining Process: Overview and Recent Advances. *J. Electrochem. Soc.* **2019**, *166*, E293. [CrossRef]
7. Yang, T.; Zeng, Y.B.; Hang, Y.S. Workpiece reciprocating movement aided Wire electrochemical machining using a tube electrode with an array of holes. *J. Mater. Process. Technol.* **2019**, *271*, 634–644. [CrossRef]
8. Yang, T.; Zeng, Y.B.; Sang, Y.M.; Li, S.Y. Effect of structural parameters of array of holes in the tube electrode for electrochemical cutting. *Int. J. Adv. Manuf. Technol.* **2020**, *107*, 205–216. [CrossRef]
9. Xu, C.C.; Fang, X.L.; Han, Z.; Zhu, D. Wire electrochemical machining with pulsating radial electrolyte supply and preparation of its tube electrode with micro-holes. *Appl. Sci.* **2020**, *10*, 331. [CrossRef]
10. Huang, G.Y.; Wan, Z.P.; Yang, S.; Li, Q.Y.; Zhong, G.D.; Wang, B.; Liu, Z.Y. Mechanism investigation of micro-drill fracture in PCB large aspect ratio micro-hole drilling. *J. Mater. Process. Technol.* **2023**, *316*, 117962. [CrossRef]
11. Mittal, R.K.; Yadav, S.; Singh, R.K. Mechanistic Force and Burr Modeling in High-speed Microdrilling of Ti6Al4V. *Proc. CIRP* **2017**, *58*, 329–334. [CrossRef]
12. Wei, T.; Sun, S.; Zhang, F.; Wang, X.; Wang, P.; Liu, X.; Wang, Q. A review on laser drilling optimization technique: Parameters, methods, and physical-field assistance. *Int. J. Adv. Manuf. Technol.* **2024**, *131*, 5691–5710. [CrossRef]
13. Dong, S.L.; Wang, Z.L.; Wang, Y.K.; Zhang, J. Micro-EDM drilling of high aspect ratio micro-holes and in situ surface improvement in C17200 beryllium copper alloy. *J. Alloys Compd.* **2017**, *727*, 1157–1164. [CrossRef]
14. Thao, O.; Joshi, S.S. Analysis of heat affected zone in the micro-electric discharge machining. *Int. J. Manuf. Technol. Manag.* **2008**, *13*, 201–213. [CrossRef]
15. Saxena, K.K.; Qian, J.; Reynaerts, D. A review on process capabilities of electrochemical micromachining and its hybrid variants. *Int. J. Mach. Tool. Manuf.* **2018**, *127*, 28–56. [CrossRef]

16. Sen, M.H.; Shan, H.S. A review of electrochemical macro- to micro-hole drilling processes. *Int. J. Mach. Tool. Manuf.* **2005**, *45*, 137–152. [CrossRef]
17. Liu, Y.; Huang, S.F. Experimental study on electrochemical drilling of micro holes with high aspect ratio. *Adv. Mater. Res.* **2014**, *941–944*, 1952–1955. [CrossRef]
18. Fan, Z.W.; Hourng, L.W. Electrochemical micro-drilling of deep holes by rotational cathode tools. *Int. J. Adv. Manuf. Technol.* **2011**, *52*, 555–563. [CrossRef]
19. Tsui, H.P.; Hung, J.C.; You, J.C.; Yan, B.H. Improvement of electrochemical microdrilling accuracy using helical tool. *Mater. Manuf. Process.* **2008**, *23*, 499–505. [CrossRef]
20. Liu, Y.; Li, M.; Niu, J.; Lu, S.; Jiang, Y. Fabrication of taper free micro-holes utilizing a combined rotating helical electrode and short voltage pulse by ECM. *Micromachines* **2019**, *10*, 28. [CrossRef]
21. Wang, M.; Zhang, Y.; He, Z.; Peng, W.; Micro-Hole, D. Deep micro-hole Fabrication in EMM on stainless steel using disk micro-tool assisted by ultrasonic vibration. *J. Mater. Process. Technol.* **2016**, *229*, 475–483. [CrossRef]

Disclaimer/Publisher's Note: The statements, opinions and data contained in all publications are solely those of the individual author(s) and contributor(s) and not of MDPI and/or the editor(s). MDPI and/or the editor(s) disclaim responsibility for any injury to people or property resulting from any ideas, methods, instructions or products referred to in the content.

micromachines

Article

Understanding the Processing Quality Problem for Cutting Ceramic Materials Using the Thermal-Controlled Fracture Method Induced by a Single-Surface Heat Source

Xiaoliang Cheng [1,*], Zhenzhen Cui [2,*], Junwen Chen [1], Yang Wang [3] and Lijun Yang [3]

1 School of Mechanical Engineering, Hubei University of Automotive Technology, Shiyan 442002, China; chenjw@huat.edu.cn
2 School of Automobile and Transportation, Wuxi Institute of Technology, Wuxi 210044, China
3 School of Mechatronics Engineering, Harbin Institute of Technology, Harbin 150001, China; wyyh@hit.edu.cn (Y.W.); yljtj@126.com (L.Y.)
* Correspondence: cxl1987@huat.edu.cn (X.C.); pursue2012@126.com (Z.C.)

Abstract: The thermal-controlled fracture method has been increasingly focused upon in the high-quality cutting of advanced ceramic materials due to its excellent characteristics. The successful application of this method in cutting ceramics mainly depends on the volumetric heating effect. However, most ceramics are treated using the surface heating mode. For the surface heating mode, the processing quality, including fracture trajectory and fracture quality, is far lower than the industrial application standards. This work was conducted to reveal the mechanism of this processing quality. Experiments involving cutting ceramics in single-surface heating mode indicate that the fracture trajectories of the upper and lower surfaces display a significant inconsistency, and the fracture quality is worse than that using the dual-surface heating mode. A cutting model was established to calculate the thermal stress distribution and to simulate the crack-propagation behaviors. The simulation results show good agreement with the experiment and provide the stress distribution, and are used to understand the reason for the processing quality problem. The mechanism of the trajectory deviation and uneven distribution of the fracture quality is revealed based on the simulation and calculation results. This study helps provide a deep understanding of the processing problems arising from this method and thus helps to innovate high-quality processing methods in this field.

Keywords: ceramic materials; thermal-controlled fracture method; surface heating mode; processing problem; trajectory deviation; simulation; uneven distribution; fracture quality

1. Introduction

Ceramics have been increasingly used in frontier domains such as aerospace and advanced chips due to their properties of high hardness, low thermal expansion, and excellent stability [1]. High-quality cutting technology is the foundation for preparing qualified parts from ceramic materials. However, the high hardness of ceramic materials can cause severe wear of tools in contact force cutting mode, as well as serious damage to the surface and subsurface of the machined workpiece [2–5]. Laser cutting methods can also cause the formation of heat-affected zones in the kerf, which can weaken the workpiece [6,7]. This severe damage induced by conventional cutting methods at the fracture surface requires complex subsequent processing to achieve qualified parts, causing a significant increase in manufacturing costs, which hinders the widespread application of advanced ceramic materials in industrial fields [8–10].

In 1968, the thermal-controlled fracture method (TCFM) was proposed by Lumley, which utilizes thermal stress to guide the crack propagation on brittle materials to cut materials. TCFM can achieve high cutting quality at relatively low temperature (no more than 350 °C) without damage in the kerf and without material removal [11]. After decades of development, this method has been implemented in industrial production for cutting flat panel display glass [12]. The successful application of TCFM in cutting glass depends on the volumetric heating effect caused by a specific wavelength laser irradiating glass [13]. However, the research on cutting opaque ceramic materials, such as laser cutting Al_2O_3 ceramics and microwave cutting low-dielectric ceramics, indicated that the surface heating mode causes poor processing quality [14,15]. The processing quality problems in cutting ceramic materials using TCFM under surface heating conditions mainly include whether the fracture trajectory follows the ideal direction and whether the fracture quality is good enough and uniform.

Regarding fracture trajectory, Brugan used a dual-beam CO_2 continuous laser to cut Al_2O_3 ceramics with a thickness of 2.54 mm using TCFM. The results indicated that the TCFM could achieve the cutting of Al_2O_3 ceramics in surface heating mode, and induced problems of trajectory deviation [16]. lu used microwaves to cut ceramic materials coated with graphite using TCFM, and the mechanism of uncontrolled crack propagation was studied [17]. Cheng studied the crack-propagation behavior in cutting silicon wafers using TCFM via a surface heat source induced by a laser, and proposed a method of surface pre-cracks to guide crack propagation to achieve approximate linear crack propagation [18].

Regarding fracture surface quality, Ueda used TCFM to cut crystalline silicon, Al_2O_3 ceramics, Si_3N_4 ceramics, etc. The results showed that the surface roughness of the fracture surface could reach 0.7 μm, 1.3 μm, and 100 μm, respectively [19]. Cai used TCFM to cut glass/silicon/glass sandwich materials, and the surface roughness of the fracture surface of the silicon layer under the action of a surface heat source reached 1 μm [20]. Saman used TCFM to cut different kinds of ceramics and studied the stress distribution characteristics. The results indicated that if the material absorbs a laser on a single surface, the maximum tensile stress area can easily be located on the back of the material, which seriously affects the stable propagation of cracks. Saman inferred that it is difficult to obtain good processing quality using the surface heating mode due to these stress distribution characteristics [21].

The above research indicates that cutting ceramics by TCFM can achieve ideal processing quality. However, the processing quality under the surface heating mode is far lower than that in volumetric heating mode, and should be improved for industry applications. To the authors' knowledge, there has been little research to date on the mechanism inducing poor processing quality using TCFM under a surface heat source. The crack-propagation mode and material fracture mechanism under the influence of a surface heat source have not been revealed. This hinders the improvement and optimization of the cutting quality in this mode.

In this study, microwaves were used to cut ceramic materials using TCFM under a surface heat source. The influence of processing parameters on fracture trajectory and fracture surface quality was studied. A TCFM cutting model was established to simulate the fracture behaviors. Combining experimental and simulation results, the mechanism of trajectory deviation and uneven distribution of fracture quality in cutting ceramic materials using TCFM under a surface heat source was revealed. This work helps to understand the mechanism of trajectory deviation and poor fracture surface quality in the cutting of ceramic materials using TCFM, and thus helps to put forward methods for improving the cutting quality under a surface heat source.

2. Experiment

2.1. Experimental Method

Al_2O_3 ceramic is opaque and does not absorb a laser, thus making it difficult to apply a volumetric heat mode. Because of its low dielectric property, it is also difficult for microwaves to interact with this ceramic. In this work, graphite with high permittivity was used to coat one surface of the Al_2O_3 ceramic to absorb microwaves to generate a surface heat source. Figure 1 shows the principle of producing the surface heating mode using the coating material induced by microwaves. The unabsorbed microwaves pass the ceramic body and enter the environment through the lower surface. The uneven heat distribution generated by the high electromagnetic loss on the coating material produces a thermal stress field. The cutting of the ceramic is realized by moving the thermal stress to guide crack propagation.

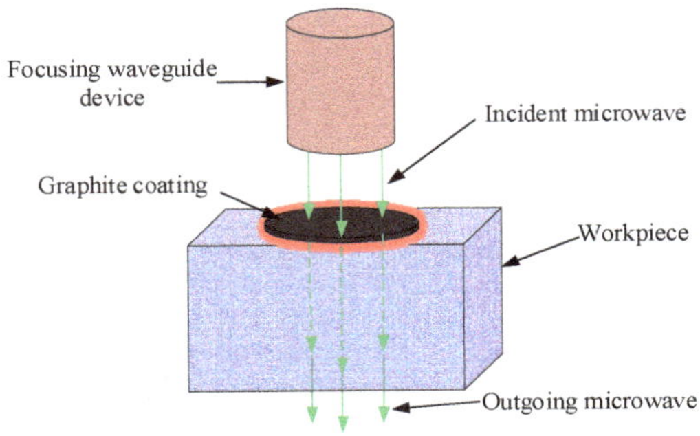

Figure 1. Schematic diagram of surface heating mode induced by microwaves.

A pre-crack was carved on the workpiece using a diamond wire with a diameter of 0.25 mm, which produced a stress amplification effect. The pre-crack with a depth of 0.1 mm is the start point of crack propagation. When the tensile stress exceeds the fracture limit at the pre-crack, the crack system reaches the propagation condition. When the waveguide is given an appropriate moving speed relative to the workpiece, the crack propagates forward and realizes cutting.

2.2. Experimental Material and Apparatus

The experiment was conducted on a microwave cutting machine. Figure 2 shows the schematic of the microwave cutting machine in TCFM. The microwave with a specific frequency is produced from the microwave source. Then, it is modulated by the microwave guide into a focusing apparatus. Finally, the focusing apparatus outputs the microwave, which is suitable for heating the material from the inner conductor. The inner conductor is surrounded by the outer shield to focus the microwave to achieve higher energy density. The coating surface of the workpiece is just below the inner conductor. The NC motion device with the x-y-z direction is integrated into the cutting machine to realize cutting movement. The z-axis is used to adjust the distance between the inner conductor and the workpiece. The x-axis is used to realize cutting movement, and the y-axis is used to center the heat source and the workpiece. This cutting device is produced by Nanjing Huiyan Microwave Equipment Company in China. It is used to cut brittle material and is equipped with safety facilities such as a filter screen to ensure safety. The microwave frequency of this device is 2.45 GHz and the maximum output power is 1500 W.

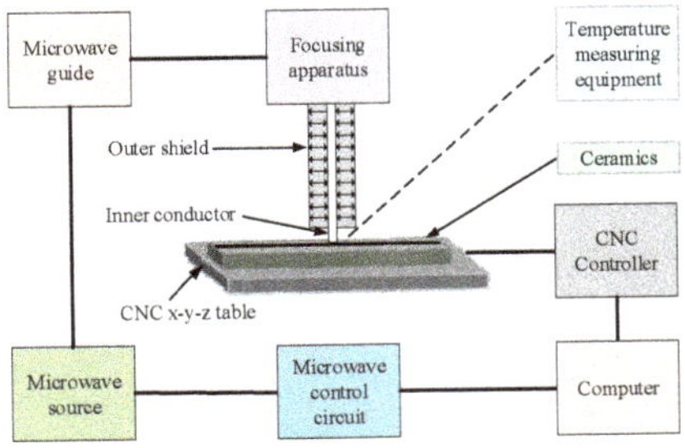

Figure 2. Schematic of microwave cutting machine using the thermal-controlled fracture method.

Figure 3 shows the Al_2O_3 ceramic plate used in this study. Graphite powder with a thickness of 0.1 mm and a width of 1 mm is coated on the expected cutting path. The plate is 100 mm × 100 mm in size. The graphite powder with a micron-size (mesh of 8000 and particle size of 1.6 μm) is mixed with alcohol. The mixture is sprayed evenly by an electric sprayer whose power comes from an air pump. As shown, a pre-crack is made on the end of the plate.

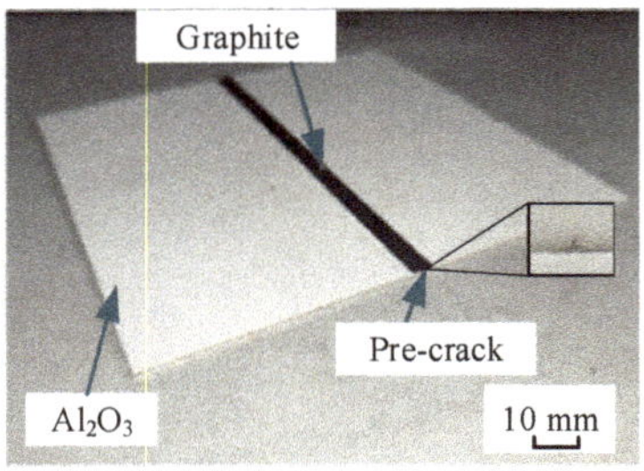

Figure 3. Al_2O_3 ceramic plate coated with graphite.

The microwave power and scanning speed were used in the experiments as processing parameters. Table 1 shows the variation range of these processing parameters. The crack propagation of the initial segment was observed by a Stemi 305 optical microscope, which was produced by CARI ZEISS in Oberkochen of Germany and can magnify objects clearly by more than 120 times.

Table 1. Variation range for processing parameters using TCFM by a surface heat source induced by microwaves.

Test Group No.	Microwave Power (W)	Scanning Speed (mm/s)
NO.1	600–1200	2.0–3.5
NO.2	900–1500	0.5–2.0
NO.3	1200–1500	0.3–0.6

To investigate the distribution characteristics of the fracture surface quality, some measurement sites were set on the section. Figure 4 shows the location of the measurement sites in the experiments. A plane on the surface heat source in workpiece was selected as the reference. Three depth positions, which are 0.2 mm, 0.5 mm, and 0.8 mm from the reference plane in the direction of the thickness, were selected to measure the surface roughness. To achieve stable fracture quality and avoid the influence of unstable boundary conditions, four horizontal inspection positions, namely 40 mm, 50 mm, 60 mm, and 70 mm from the inlet along the cutting direction, were selected. To ensure the value of the experimental results, each experiment was repeated four times, and the average of the results was taken.

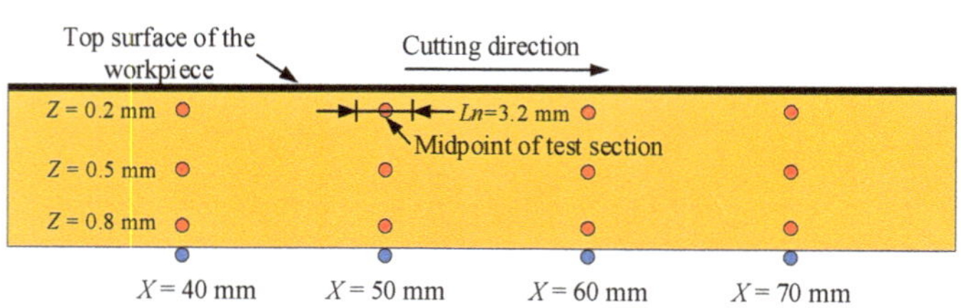

Figure 4. Position of the surface roughness test.

3. Results

3.1. Fracture Trajectory

Figure 5 shows the fracture trajectories at each surface of the Al_2O_3 ceramic plates after cutting at a microwave power of 1000 W and a scanning speed of 3 mm/s. It can be shown that the fracture trajectory approaches a straight line when cutting ceramics with a surface heat source. Comparing the processing effect of graphite in this study and clay as the microwave-absorbing material in the literature [22], it is notable that pure graphite micron powder can achieve better processing quality.

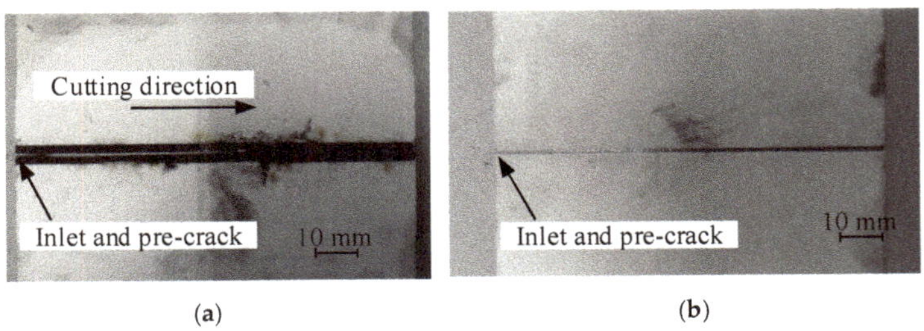

Figure 5. Experimental results of fracture trajectories of surface-heat-source cutting Al_2O_3 ceramic under using thermal-controlled crack propagation induced by microwaves: (**a**) the upper surface; (**b**) the lower surface.

Figure 6 shows a photo magnified with an optical microscope of the fracture trajectory on the two surfaces of the workpiece. It can be shown that the fracture trajectories of the upper and lower surfaces display significant inconsistencies from a microscopic perspective.

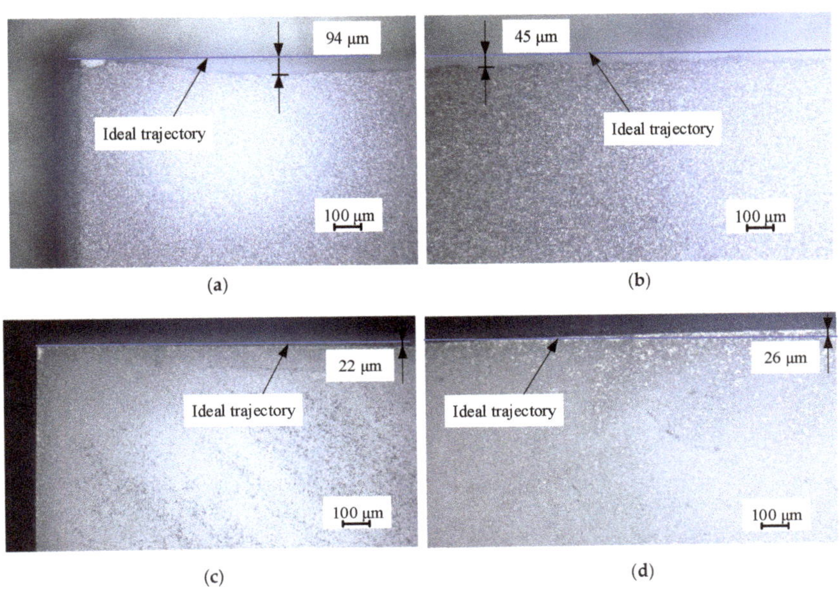

Figure 6. Optical micrograph of deviation in initial crack propagation: (**a**) inlet of upper surface; (**b**) extension of upper surface; (**c**) outlet of lower surface; (**d**) initial extension of lower surface.

The effect of processing parameters on fracture trajectory was studied. Figure 7 shows the relationship between the microwave power and the trajectory deviation of the initial segment. It shows the effect of the microwave power on the initial trajectory deviation length *Le* and maximum offset *De* of the upper and lower surfaces at a cutting speed of 3 mm/s and a microwave power of 1000 W, 1100 W, 1200 W, and 1300 W. As is shown, the deviation length *Le* and maximum offset *De* both increase as the microwave power increases, and the *De* increases more significantly. Therefore, it is necessary to choose an appropriate microwave power in order to reduce the trajectory deviation.

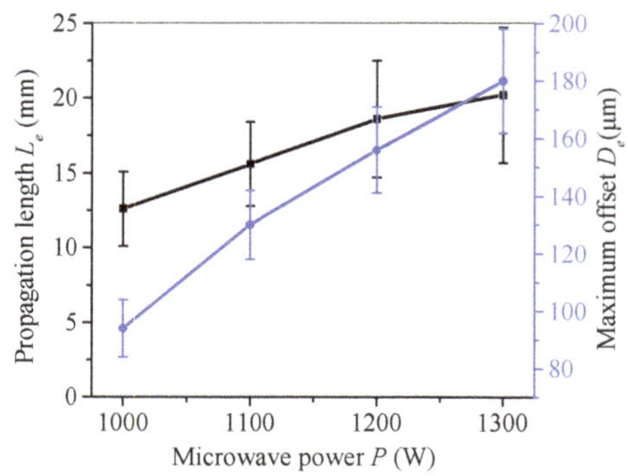

Figure 7. Effect of microwave power on the offset propagation length and the maximum offset extension.

3.2. Fracture Surface Quality

Figure 8 show the fracture surface of cutting Al_2O_3 ceramic plates at a microwave power of 1000 W and a scanning speed of 3 mm/s. As is shown, there is obvious processing damage at the entrance due to the process of prefabricating cracks using a diamond wire saw. It is notable that the fracture quality of the middle and outlet sections is better than that at the entrance, and the middle section is the best. There is a small amount of edge damage at the outlet.

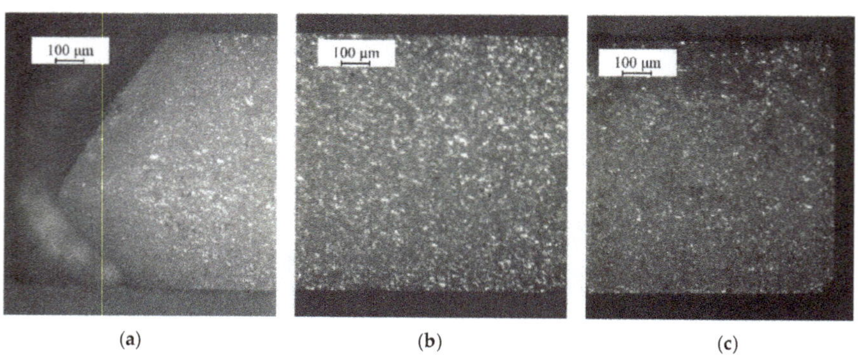

Figure 8. Experimental results of fracture surface of Al_2O_3 ceramics under the action of thermally induced crack propagation using a surface heat source:(**a**) the cutting inlet; (**b**) the middle segment; (**c**) the cutting outlet.

Figure 9 shows the variation in the arithmetic mean deviation of surface roughness *Ra* at the middle thickness of the workpiece along the cutting direction, ranging from 1.5 mm to 4.7 mm, 30 mm to 33.2 mm, 60 mm to 63.2 mm, and 95 mm to 98.2 mm, when the microwave power is 1000 W and the cutting speed is 3 mm/s. The sampling length is 0.8 mm, and the evaluation length is 3.2 mm. It is notable that the surface roughness *Ra* of the initial scanning segment is 1 to 2 orders of magnitude higher than that of the middle and end segments, which corresponds to the inconsistent propagation of surface cracks on the upper and lower surfaces.

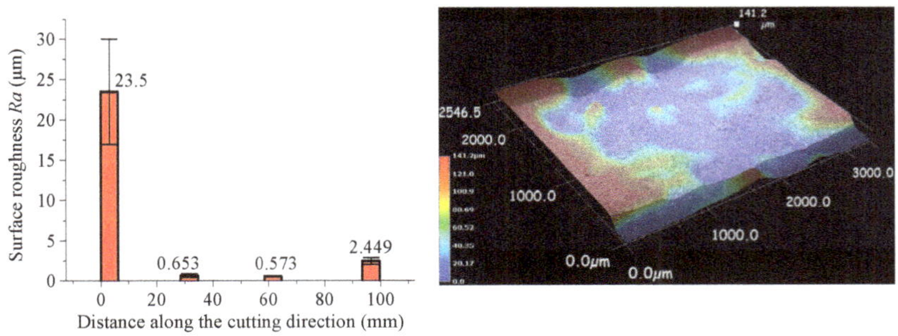

(**a**) (**b**)

Figure 9. Surface roughness and fracture surface along the scanning direction at the middle depth: (**a**) surface roughness; (**b**) 3D figure of fracture surface.

Figure 10 shows the variation in surface roughness *Ra* in the thickness direction of the workpiece. As is shown, the average *Ra* at most points is between 0.2 μm to 1 μm. It is notable that there is a significant difference in *Ra* along the depth direction at various cutting positions without obvious regularity.

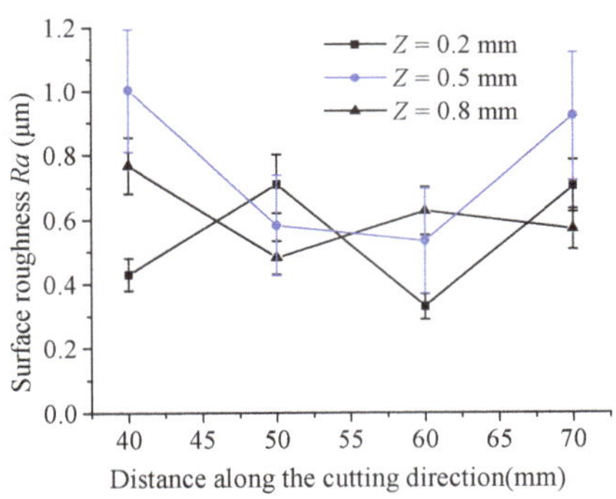

Figure 10. Surface roughness distribution at different positions of the material at the middle segment.

Figure 11 shows the effect of microwave power on *Ra* under the condition of a cutting speed of 3 mm/s at several positions along the cutting direction. It indicates that the surface roughness of each position increases as the microwave power increases. The microwave power has a significant impact on the surface roughness of the initial segment. When the maximum microwave power is 1300 W, the surface roughness of both the initial and final segments will increase significantly. Therefore, it is necessary to choose an appropriate microwave power when using a surface heat source for cutting.

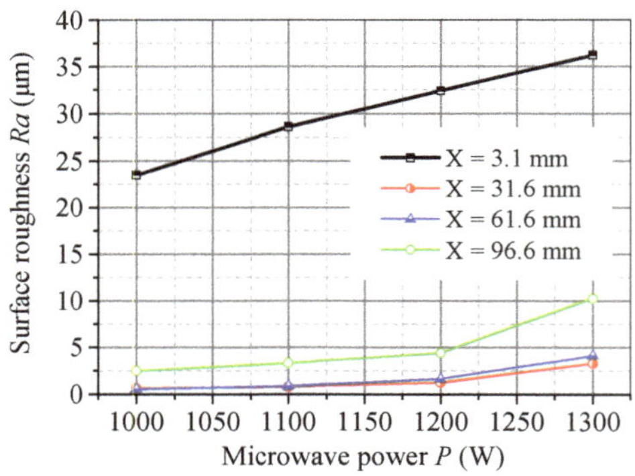

Figure 11. Effect of microwave power on the surface roughness along the scanning direction at different positions in the middle depth of the material.

4. Discussion

4.1. Finite Element Modeling

The main physical processes of cutting ceramic materials using TCFM are interactions between energy beams and matter, heat generation, heat transfer, thermal stress generation, and crack propagation.

Since cracks tend to propagate perpendicularly to the maximum transverse tensile stress component, which is affected by the stress intensity factor in different fracture modes, the fracture problem is transformed into determining the maximum value of the mechanical energy release rate in fracture mechanics.

The crack-propagation condition is that the tensile stress at the tip of the pre-crack exceeds the fracture limits of the material. From an energy perspective, it is a process of converting elastic energy into surface energy of the fracture surface.

The cutting of ceramics materials by TCFM is a complex physical process. It is difficult to calculate the temperature, thermal stress, and crack propagation using an analytical model. A finite element modeling (FEM) technology with the aid of ABAQUS 6.14 software was used in this study to calculate these physical quantities and simulate the cutting process.

The cutting simulation was conducted for both surface and volumetric heat sources. Figure 12 shows the simulation results of crack propagation for cutting Al_2O_3 ceramics by a microwave surface heat source and the crack propagation for cutting glass with a volumetric heat source. From Figure 12a, it is notable that the crack propagates inconsistently between the upper and the lower surface of the workpiece. The crack propagation on the upper surface presents a significant folding line, while the lower surface is relatively straight. From Figure 12b, it is notable that that the crack-propagation paths on the upper and lower surfaces of the workpiece are consistent under the action of a volumetric heat source, and the crack-propagation paths tend to be straight.

Figure 13 show a comparison between the experimental and simulation results of the crack-propagation trajectory under a surface heat source. It shows that the degree of offset generated in experiments and simulations is relatively close. Therefore, the simulation of crack propagation in this study is suitable for discussing the cutting quality.

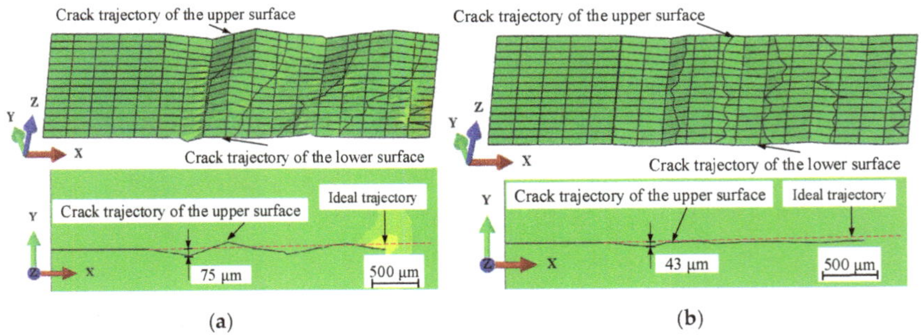

Figure 12. Simulation of initial crack-propagation path and section quality via the thermal cracking method using a surface heat source and its comparison with a volumetric heat source mode: (**a**) crack propagation induced by a surface heat source using the thermal cracking method; (**b**) crack propagation induced by a volumetric heat source using the thermal cracking method.

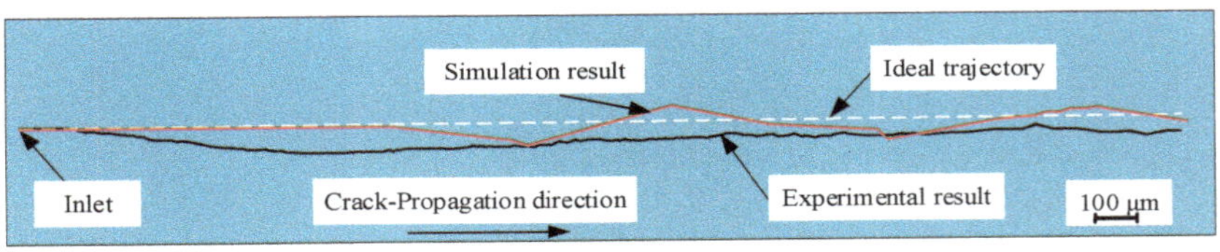

Figure 13. Comparison of experimental and simulation results of the initial crack-propagation trajectory.

4.2. Trajectory Deviation Mechanism

There are three basic modes of crack propagation and fracture, which are type I (open), type II (sliding), and type III (tear). Figure 14 shows a schematic of these three basic fracture modes. Among them, type I means that the crack only breaks under the action of tensile stress perpendicular to the fracture surface; type II means that the crack breaks only under the action of longitudinal shear stress perpendicular to the leading edge of the crack; type III means that the crack breaks only under the action of lateral shear stress parallel to the leading edge of the crack.

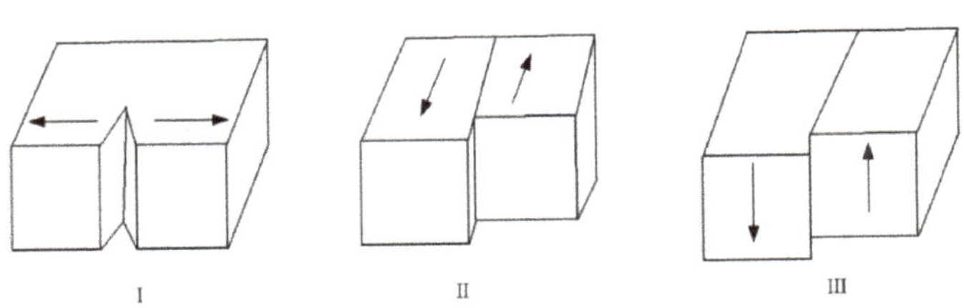

Figure 14. Three basic fracture modes.

Among the three basic fracture modes, type I is closest to the fracture of materials with high brittleness. Due to the high brittleness of the ceramic materials in this study, the fracture mode is generally considered to be type I. However, during the fracture and propagation of the actual crack, it will be affected by both the sliding and tear of the others. In this way, the crack is usually a "tilted" (type I and type II composite) fracture or "torsional" (type I and type III composite) fracture according to the actual influence of these two effects during the crack-propagation process.

The main reasons for the occurrence of offset propagation of cracks are divided into "inclined" type (type I and type II composite) fractures or "torsional" type (type I and type

III composite) fractures. Figure 15 shows the schematic diagrams of these two typical types. It adopts the same coordinate system and crack location as the simulation calculation, where the "inclined" type in Figure 16a refers to the original crack surface producing a deflection angle θ relative to the original propagation direction under the influence of transverse shear force, and an extended length dc_1 of the included angle. The "torsional" propagation in Figure 15b is caused by the influence of lateral shear force on the original crack surface, resulting in a deflection angle φ relative to the original propagation direction, and the extended length dc_2 of the included angle.

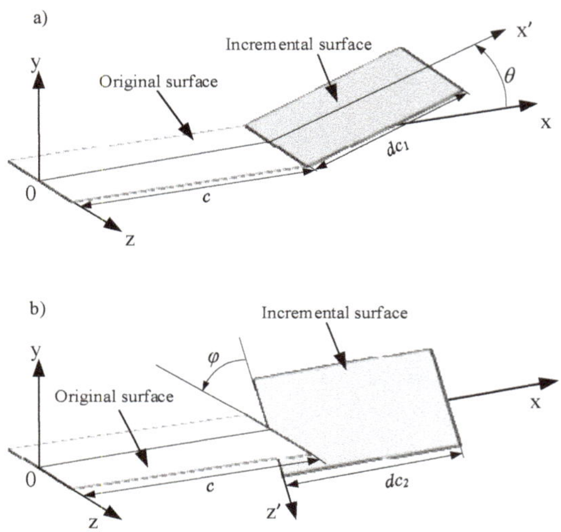

Figure 15. Two typical deviation types of crack propagation: (**a**) inclined type; (**b**) torsion type.

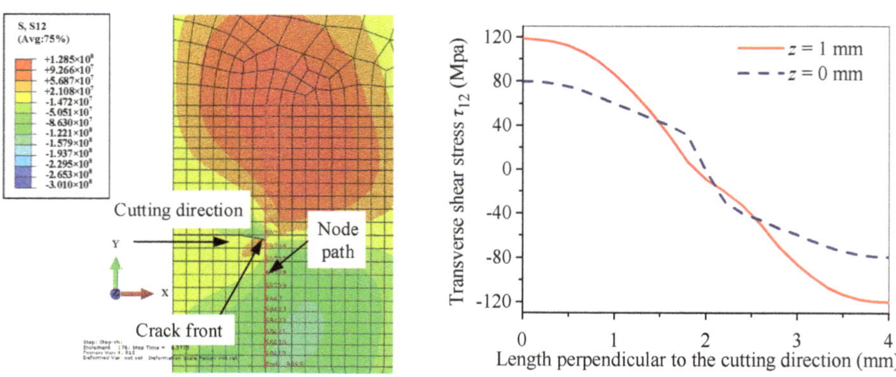

(**a**) (**b**)

Figure 16. Transverse shear stress distribution front perpendicular to the scanning direction of a surface heat source in a Al_2O_3 ceramic surface crack at t = 1.9058 s: (**a**) cloud chart of transverse shear stress distribution and node path on the upper surface of the workpiece; (**b**) transverse shear stress distribution of the crack front perpendicular to the scanning direction.

In order to study the mechanism of deviation of crack propagation in a surface heat source-induced thermal fracture, it is necessary to analyze the two kinds of shear stress near the crack that cause the migration. These are S_{12} and S_{13} in the output of the simulation results, which present the shear stress τ_{12} causing "inclined" propagation and shear stress τ_{13} causing "torsional" propagation.

Figure 16 shows the distribution of transverse shear stress τ_{12}, which is perpendicular to the cutting direction at the crack front at t = 1.9058 s. To obtain effective data, a grid segment with a length of 2 mm was symmetrically taken on the front edge of the crack perpendicular to the initial crack plane, and the center position of this segment is the ideal position for the crack propagation without deviation. The number of selected nodes is 23, and the path direction of the selected nodes is in the positive y-axis direction. As is shown in Figure 16a, the crack on the upper surface of the workpiece shifted to the right. Figure 16b shows that τ_{12} of the upper surface changes more significantly than the lower surface, which is caused by the surface heat source being located on the upper surface of the workpiece. It is notable that the intersection of the τ_{12} curve and the 0 MPa line is close to the right side of the symmetry line. According to the theory of fracture mechanics, the crack always tends to propagate where the shear stress is the smallest, so the upper surface crack has a deviation to the right.

Figure 17 shows the crack propagation on the lower surface of the workpiece. It indicates that the crack on the lower surface of the workpiece is returning to the ideal propagation path. This is consistent with the middle position of the intersection of the τ_{12} curve and the 0 MPa line. Figure 18 shows the τ_{12} curve on each surface of the workpiece at multiple times during the subsequent crack-propagation process. It is notable that the significant size differences of τ_{12} on the upper and lower surfaces of the workpiece result in their different effects on the tilting behavior of the cracks. This is the reason for the inconsistent crack propagation on upper and lower surfaces, as well as their different deviation behaviors in experiments.

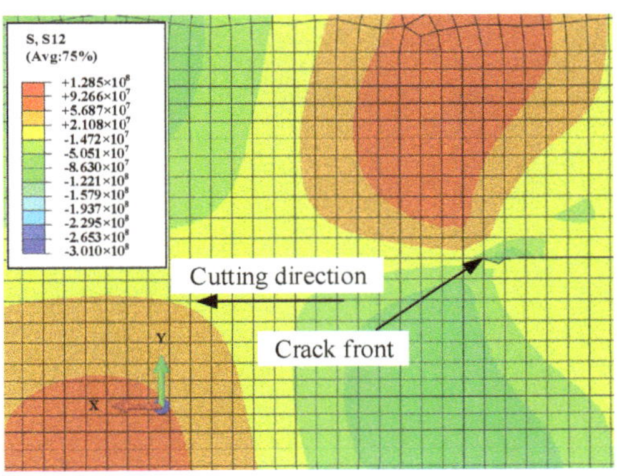

Figure 17. Crack propagation on the lower surface of the workpiece at t = 1.9058 s.

Figure 19 shows the lateral shear stress distribution near the crack front edge at t = 1.7675 s. The plane X = 1.2 mm perpendicular to the x-axis is made at the crack front edge to display the lateral shear stress distribution shown in Figure 19a. It is notable that the lateral shear stress has a significantly asymmetric distribution within a range of 400 μm around the front edge of the crack. The influence of this lateral shear stress distribution on the formation near the crack front leads to torsional propagation, as shown in Figure 19b. Due to the direct heating effect on the upper surface of the workpiece, the twisting behavior near the upper surface is more pronounced. The trajectory deviation occurs under the combined action of lateral shear stress and transverse shear stress in cutting ceramics with a surface heat source.

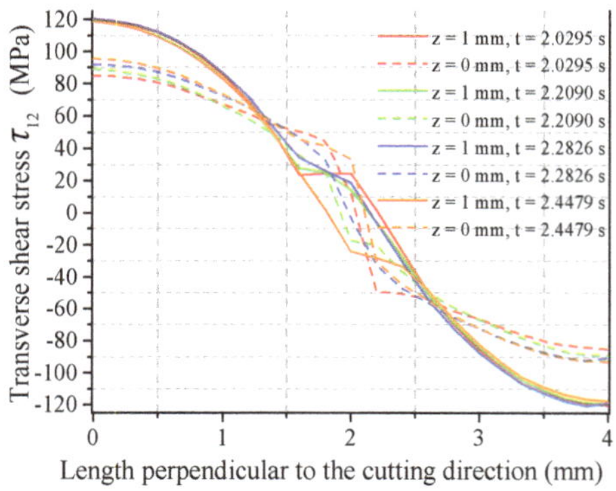

Figure 18. Transverse shear stress distribution at the upper and lower surfaces of the crack front at different times.

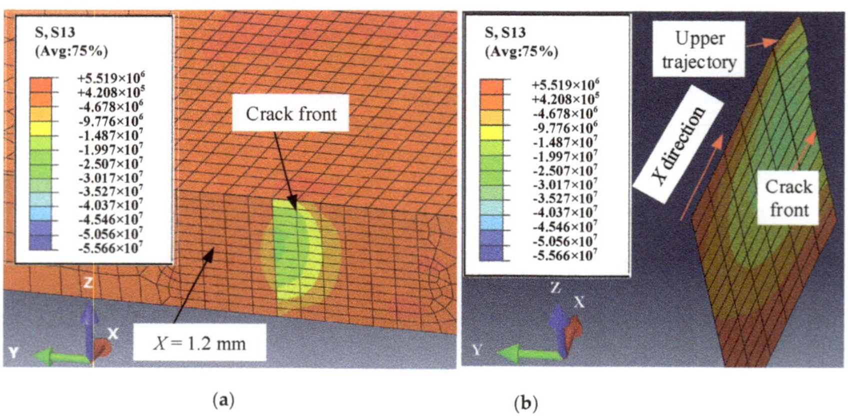

Figure 19. Lateral shear stress distribution near the front edge of the crack when the crack propagates with torsion at t = 1.7675 s: (**a**) distribution of lateral shear stress near the crack front; (**b**) torsional propagation of the crack.

4.3. Uneven Mechanism of Fracture Surface Quality

The fracture surface is formed under the guidance of transverse tensile stress whose characteristic determines the mode of crack propagation in TCFM. Figure 20 shows the distribution of transverse tensile stress at the crack front and its circumferential condition when the crack is about to propagate.

It indicates that the stress gradient near the crack front is relatively large when it is about to propagate, and the transverse tensile stress is large near the surface under the condition of a surface heat source.

The gradient of transverse tensile stress along the thickness direction of the material indicates that crack propagation is severely asynchronized at different positions of the cross-section. As was reported in [18], the propagation first occurs from the root of the existing crack on the upper surface, and the crack expands to the lower surface along the thickness direction. The incentive for its subsequent expansion is that the expanding crack forms a new tip with the existing crack in the vertical direction, which has the minimum curvature radius and the most concentrated stress to guide the crack to propagate downwards. This crack-propagation mode results in the uneven distribution of the fracture quality in the cutting experiment.

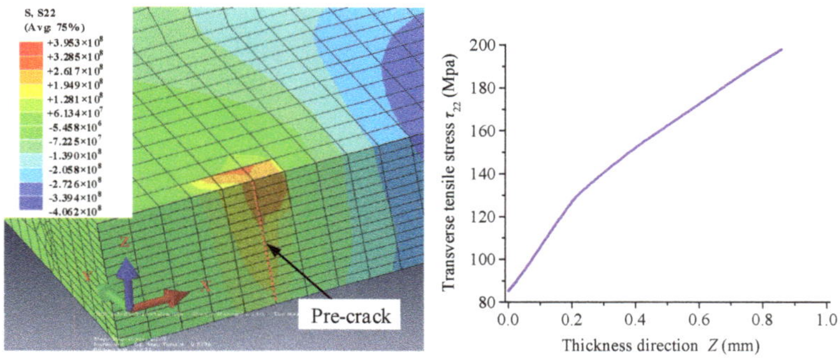

(**a**) (**b**)

Figure 20. Distribution of transverse tensile stress at the crack front at the moment of propagation: (**a**) circumferential condition of crack front edge; (**b**) crack front edge.

To improve the large stress gradient in single-sided surface heating mode, a dual-sided surface heating method was proposed by the authors of this article in the literature [23]. Figure 21 shows the schematic diagram of heating ceramics using a microwave dual-surface heat source. The lower surface of the workpiece is coated by graphite, which is used to absorb the underutilized microwaves and forms a lower surface heat source.

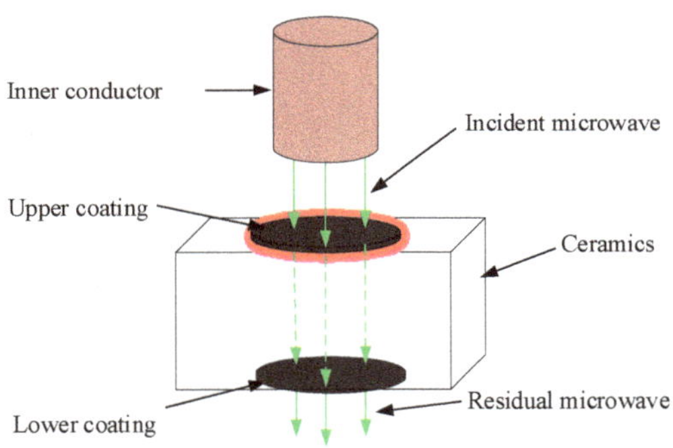

Figure 21. Schematic diagram of heating ceramics using a microwave dual-surface heat source.

Figure 22 shows the transverse tensile stress at the crack front along the direction of material thickness at the moment of propagation under the action of a dual-surface heat source at different times. As is shown, the maximum transverse tensile stress zone under the action of a dual-surface heat source is located at the middle section of the crack root. This indicates that the difference in the maximum transverse tensile stress zone and its gradient at the crack front edge under different heat sources determines their different crack-propagation modes and their fracture quality. According to this study, the distribution of the transverse tensile stress at the crack front edge determines the distribution of fracture quality.

Cutting experiments under a dual-sided surface heat source were conducted. Figure 23 shows the surface roughness distribution at the middle segment at the condition of 600 W of microwave power and a cutting speed of 3 mm/s under dual-surface heat mode. As is shown, the fracture quality and its distribution under the action of a dual-sided surface are significantly better, and the distribution is more uniform than that of a single-surface heat source. The middle depth of the section has the best fracture quality, which has good agreement with that revealed in reference [19].

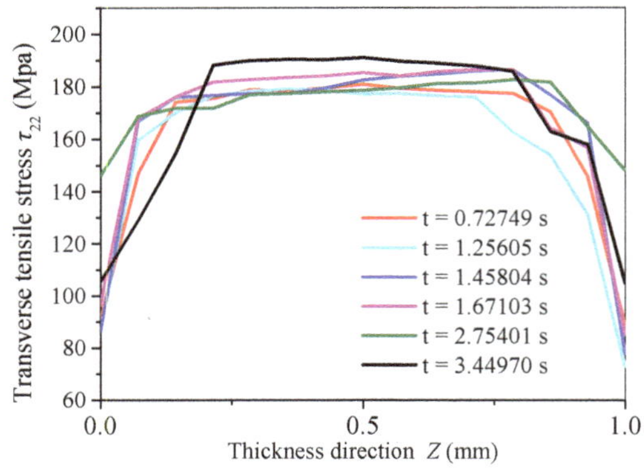

Figure 22. Distribution of transverse tensile stress along the thickness direction at the crack front at the moment of propagation under the action of a dual-surface heat source at different times.

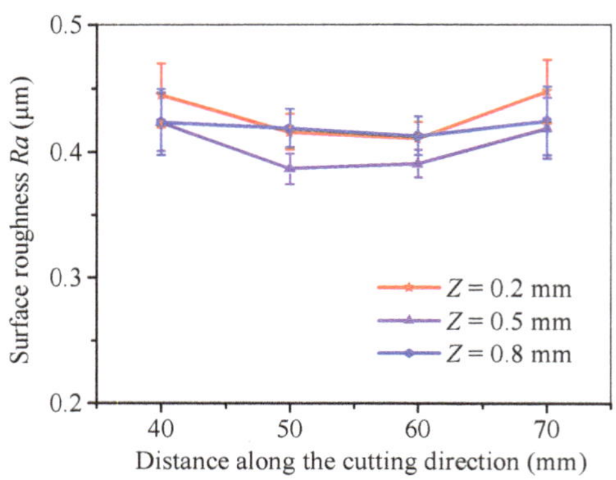

Figure 23. Surface roughness distribution at the middle segment under dual-surface heat mode.

5. Conclusions

In this work, experiments and simulations of cutting ceramics using the thermal-controlled fracture method with a surface heat source were conducted to reveal the mechanism of trajectory deviation and uneven distribution of fracture quality. Some conclusions can be made from this study, as follows.

A finite element model was established to calculate the physical quantities and simulate the cutting process in the thermal-controlled fracture method in surface heating mode. The crack-propagation results of the simulation were consistent with the trend of the cutting path in the experiment. The modeling results could be used to reveal the mechanism of trajectory deviation and uneven distribution of fracture quality for this method.

The trajectory deviation occurs under the combined action of lateral shear stress and transverse shear stress in cutting ceramics with a surface heat source.

The significant differences in τ_{12} on the upper and lower surfaces of the workpiece result in their different effects on the tilting behavior of the cracks and the inconsistent crack propagation. This is the reason for the trajectory deviation under a surface heat source.

The significantly asymmetric distribution of lateral shear stress around the front edge of the crack leads to torsional propagation under a surface heat source.

The distribution of the transverse tensile stress along the thickness direction of the workpiece shows a positive correlation with the uneven distribution of fracture quality in the single-surface heating mode, in contrast with the dual-sided surface and volumetric heat source cutting method.

Future research could focus on innovative techniques that could reduce the gradient of transverse tensile stress in the thickness direction and correct the asymmetric distribution of the two types of shear stress.

Author Contributions: Methodology, Y.W. and Z.C.; Software, X.C.; Validation, J.C.; Formal analysis, X.C.; Investigation, J.C.; Resources, Z.C.; Data curation, L.Y.; Writing—original draft, X.C.; Writing—review and editing, Z.C.; Supervision, Z.C., X.C. and J.C.; Project administration, X.C. and L.Y.; Funding acquisition, X.C. All authors have read and agreed to the published version of the manuscript.

Funding: This work was supported by the PhD research startup foundation of Hubei University of Automotive Technology (BK202403, BK202330) and Natural Science Research of Jiangsu Higher Education Institutions of China (No. 23KJB430038).

Data Availability Statement: Data available on request from the authors.

Acknowledgments: We thank Harbin Institute of Technology for its support in processing equipment, processing testing, and modeling of this process.

Conflicts of Interest: The authors declare no conflicts of interest.

References

1. Sun, J.; Yu, H.; Zeng, D.; Shen, P. Wire–powder–arc additive manufacturing: A viable strategy to fabricate carbide ceramic/aluminum alloy multi-material structures. *Addit. Manuf.* **2022**, *51*, 102637. [CrossRef]
2. Rakshit, R.; Das, A. A review on cutting of industrial ceramic materials. *Precis. Eng.* **2019**, *59*, 90–109. [CrossRef]
3. Lin, H.; Zhou, M.; Wang, H.; Bai, S. Investigation of Cutting Force and the Material Removal Mechanism in the Ultrasonic Vibration-Assisted Scratching of 2D-SiCf/SiC Composites. *Micromachines* **2023**, *14*, 1350. [CrossRef]
4. Wu, Q.; Zhou, X.; Pan, X. Cutting tool wear monitoring in milling processes by integrating deep residual convolution network and gated recurrent unit with an attention mechanism. *Proc. Inst. Mech. Eng. Part B* **2023**, *237*, 1171–1181. [CrossRef]
5. Chen, M.; Zhang, S.; Wu, Y.; Wang, H. Build an accurate 3D geometrical model of a soft knife profile of abrasive water jet. *The Int. J. Adv. Manuf. Tech.* **2023**, *129*, 2475–2489. [CrossRef]
6. Yang, L.J.; Ding, Y.; Cheng, B.; Mohammed, A.; Wang, Y. Numerical simulation and experimental research on reduction of taper and HAZ during laser drilling using moving focal point. *Int. J. Mach. Tool. Manu.* **2017**, *91*, 1171–1180. [CrossRef]
7. Li, C.; Hu, Y.; Zhang, F.; Geng, Y.; Meng, B. Molecular dynamics simulation of laser assisted grinding of GaN crystals. *Int. J. Mech. Sci.* **2023**, *239*, 107856. [CrossRef]
8. Li, C.; Piao, Y.; Zhang, F.; Zhang, Y.; Hu, Y.; Wang, Y. Understand anisotropy dependence of damage evolution and material removal during nanoscratch of MgF_2 single crystals. *Int. J. Extreme Manuf.* **2023**, *5*, 015101. [CrossRef]
9. Li, C.; Piao, Y.; Meng, B.; Hu, Y.; Li, L.; Zhang, F. Phase transition and plastic deformation mechanisms induced by self-rotating grinding of GaN single crystals. *Int. J. Mach. Tool. Manuf.* **2022**, *172*, 103827. [CrossRef]
10. Lin, W.; Yu, D.; Zhang, C.; Zhang, S.; Tian, Y.; Liu, S.; Luo, M. Multi-objective optimization of machining parameters in multi-pass turning operations for low-carbon manufacturing. *Proc. Inst. Mech. Eng. Part B* **2017**, *231*, 2372–2383. [CrossRef]
11. Lumley, R.M. Controlled Separation of Brittle Materials Using a Laser. *Am. Ceram. Soc. Bull.* **1968**, *48*, 850–854.
12. Kim, K.R.; Kim, J.H.; Farson, D.F.; Choi, H.W.; Kim, K.H. Hybrid laser cutting for flat panel display glass. *Jpn. J. Appl. Phys.* **2008**, *47*, 6978. [CrossRef]
13. Zhao, C.; Zhang, H.; Wang, Y. Semiconductor laser asymmetry cutting glass with laser induced thermal-crack propagation. *Opt. Laser. Eng.* **2014**, *63*, 43–52. [CrossRef]
14. Tsai, C.; Chen, H. Laser cutting of thick ceramic substrates by controlled fracture technique. *J. Mater. Process. Technol.* **2003**, *136*, 166–173. [CrossRef]
15. Wang, H.; Zhang, H.; Wang, Y. Splitting of glass and SiC ceramic sheets using controlled fracture technique with elliptic microwave spot. *Ceram. Int.* **2016**, *43*, 1669–1676. [CrossRef]
16. Brugan, P.; Cai, G.; Akarapu, R.; Segall, A.E. Controlled-fracture of prescored alumina ceramics using simultaneous CO_2 lasers. *J. Laser. Appl.* **2006**, *18*, 236–241. [CrossRef]
17. Lu, Y.; He, Z.; Xu, J.; Wang, Y.; Yang, L. A novel method by microwave cutting ceramics based on thermal crack and trajectory control. *Int. J. Adv. Manuf. Tech.* **2023**, *125*, 3853–3867. [CrossRef]
18. Cheng, X.; Yang, L.; Wang, M.; Cai, Y.; Wang, Y.; Ren, Z. Laser beam induced thermal-crack propagation for asymmetric linear cutting of silicon wafer. *Opt. Laser. Technol.* **2019**, *120*, 105765. [CrossRef]
19. Ueda, T.; Yamada, K.; Oiso, K.; Hosokawa, A. Thermal Stress Cleaving of Brittle Materials by Laser Beam. *Cirp. Ann. Manuf. Technol.* **2002**, *51*, 149–152. [CrossRef]
20. Cai, Y.; Wang, M.; Zhang, H.; Yang, L.; Fu, X.; Wang, Y. Laser cutting sandwich structure glass–silicon–glass wafer with laser induced thermal—Crack propagation. *Opt. Laser. Technol.* **2017**, *93*, 49–59. [CrossRef]

21. Saman, A.M.; Furumoto, T. Evaluation of Separating Process for Different Materials by Thermal Stress Cleaving Technique. *Int. J. Precis. Eng.* **2018**, *7*, 158–161. [CrossRef]
22. Wang, H.; Zhang, H.; Wang, Y.; Wang, M. Thermal controlled fracture of Al_2O_3 substrate by inducing microwave discharge in graphite coat. *Ceram Int.* **2019**, *45*, 6149–6159. [CrossRef]
23. Cheng, X.; He, Z.; Wang, H.; Wang, Y. Splitting Opaque, Brittle Materials with Dual-Sided Thermal Stress Using Thermal-Controlled Fracture Method by Microwave. *Crystals* **2022**, *12*, 801. [CrossRef]

Disclaimer/Publisher's Note: The statements, opinions and data contained in all publications are solely those of the individual author(s) and contributor(s) and not of MDPI and/or the editor(s). MDPI and/or the editor(s) disclaim responsibility for any injury to people or property resulting from any ideas, methods, instructions or products referred to in the content.

micromachines

MDPI

Article

Experimental Study on Chemical–Mechanical Synergistic Preparation for Cemented Carbide Insert Cutting Edge

Changjiang Qin *, Jian Pan, Lei Guo, Chi Zhang, Wanli Chen, Zihua Hu, Shengqiang Jiang, Xiaogao Chen and Meijiao Mao

School of Mechanical Engineering and Mechanics, Xiangtan University, Xiangtan 411105, China; panjian802@163.com (J.P.); 13551757978@163.com (L.G.); m19976601099@163.com (C.Z.); a15623412023@163.com (W.C.); iamtxtu@163.com (Z.H.); jsqcx@126.com (S.J.); 17769498801@163.com (X.C.); maomj2000@163.com (M.M.)

* Correspondence: qincjxtu@163.com; Tel.: +86-0731-58292209

Abstract: Typical edge defects in the edge region of a new cemented carbide insert without edge preparation include burrs, poor surface quality, micro-breakages, and irregularities along the edge. To address the problems in new cemented carbide inserts without edge preparations, a chemical–mechanical synergistic preparation (CMSP) method for the cemented carbide insert cutting edge was proposed. Firstly, the CMSP device for the insert cutting edge was constructed. Then, the polishing slurry of the CMSP for the insert cutting edge was optimized using the Taguchi method combined with a grey relation analysis and fuzzy inference. Finally, orthogonal experiments, the Taguchi method, and analysis of variance (ANOVA) were used to investigate the effect of the polishing plate's rotational speed, swing angle, and input frequency of the controller on the edge preparation process, and the parameters were optimized. The results showed that the best parameter combination for the polishing slurry for the cemented carbide inserts was the mass concentration of the abrasive particle of 10 wt%, the mass concentration of the oxidant of 10 wt%, the mass concentration of the dispersant of 2 wt%, and the pH of 8. The CMSP process parameter combination for the linear edge had the polishing plate's rotational speed of 90 rpm, the swing angle of 6°, and the input frequency of the controller of 5000 Hz. The optimum CMSP process parameter combination for the circular edge had the polishing plate's rotational speed of 90 rpm, the swing angle of 6°, and the input frequency of the controller of 7000 Hz. The polishing plate's rotational speed had the most significant impact on the edge preparation process, followed by the swing angle, and the effect of the input frequency of the controller was the smallest. This study demonstrated that CMSP is a potential way to treat the cemented carbide insert cutting edge in a tool enterprise.

Keywords: cemented carbide insert; chemical–mechanical synergistic preparation (CMSP); cutting edge; material removal rate (MRR); surface roughness

1. Introduction

Due to significant hardness, wear resistance, strength, and toughness, cemented carbide inserts have broad applications in cutting difficult-to-machine materials such as stainless steel, titanium alloys, and superalloys [1]. To meet the precision machining requirements of high contour accuracy and high surface quality for the parts, stable ultra-precision machine tools, process parameters, and machining environments are required [2,3], but a high-quality cemented carbide insert is one of the key technologies. In the machining process, the cutting contact region between a carbide insert and a part includes the rake face, the flank face, and the circular arc or linear region of the cutting edge. The cutting edge is responsible for the main task of material removal. The geometrical parameters and mechanical properties of a cutting edge directly affect the stability of the machining process and the machined surface's roughness [4]. Therefore, it is required that the surface integrity of a cemented carbide insert cutting edge is good, there is no surface/sub-surface damage layer, stable cutting can be carried out, and the tool life can be improved. Therefore, production costs can be reduced and material utilization can be improved. The surface of a new, sharp cemented carbide insert without an edge preparation is machined with diamond grinding in various conditions, and there are typical edge defects in the edge region, such as micro-breakages, burrs, bad surface quality, and irregularities along the edge. To reduce and eliminate edge defects and improve the cutting performance of an insert, cutting edge preparation is carried out by tool manufacturers after the final grinding process of an insert manufacturing process is completed. Many studies have been conducted on cemented carbide insert cutting edge preparation methods. Denkena et al. [5] proposed a five-axis brushing technique for preparing a carbide insert edge and found that an insert with edge preparation could improve the machined surface quality of a part. Yussefian et al. [6] used electrical spark discharges to prepare a cemented carbide insert edge and found that an insert with edge preparation could improve tool life. Zhang et al. [7] used a wet micro-abrasive blasting method to prepare a carbide insert edge, and this method could improve tool life and cutting performance. Wang et al. [8] used a pressurized air wet abrasive jet machining method to prepare a cemented carbide insert cutting edge, and it was found that an insert with edge preparation could improve tool life and achieve better part surface quality. Wang et al. [9] used wet abrasive jet machining, brushing, and drag finishing methods to prepare cemented carbide insert edges, and they found that the residual compressive stress levels of the edges prepared with abrasive jet machining were 63% higher than those of the edges without preparation, and the surface roughness levels of the edges prepared with drag-finishing were the lowest, while the cutting performances of the edges prepared with drag finishing were also the best. Zimmermann et al. [10] used laser machining to prepare the cutting edge of a cemented carbide tool and found that the formation of residual tensile stresses and the deterioration of the mechanical properties of the surface layer of a tool could be reduced with appropriate laser-machining parameters.

To obtain a better surface quality of an insert cutting edge, some scholars have proposed a polishing method to prepare the edges of carbide inserts. Lyu et al. [11] proposed a brush-assisted shear-thickening polishing method for the preparation of a cemented carbide insert cutting edge, and this method could obtain better surface quality of a cutting edge. Shao et al. [12] employed a flexible fiber-assisted shear-thickening polishing method for the preparation of a carbide insert cutting edge and studied the effect of the polishing angle and polishing speed on the efficiency of the cutting edge preparation. This method could also obtain a better surface quality of a cutting edge.

The chemical–mechanical polishing (CMP) method is a widely recognized ultra-precision machining method with global planarization at the nanoscale [13]. This method mainly relies on the mechanical action of abrasive particles and the chemical action of oxidants in a polishing slurry to make the surface of a polished workpiece flat and smooth, which has the advantages of no surface/subsurface damage, low density of surface defects, and low cost [14]. Moreover, CMP technology has been employed for polishing the rake faces of carbide inserts, and it has been found that this method can improve the surface quality and cutting performance of an insert [15,16]. In addition, Qin et al. [17] investigated the material removal mechanism of the chemical mechanical polishing of a carbide insert and established a material removal model for the chemical mechanical polishing of a carbide insert.

Because of the advantages of CMP technology, the chemical–mechanical synergistic preparation (CMSP) of the cemented carbide insert cutting edge was proposed to improve the surface integrity of the cemented carbide insert edge. Firstly, the device for the CMSP of cemented carbide insert cutting edges was constructed. Then, the polishing slurry for the CMSP of the cemented carbide insert cutting edge was optimized using the Taguchi method combined with a grey relation analysis and fuzzy inference. Finally, orthogonal experiments, range analysis, and analysis of variance (ANOVA) were used to investigate the effects of the polishing plate's rotational speed of the swing angle and input frequency of the controller on the edge preparation process, and the edge preparation process parameters were optimized.

2. Construction of CMSP Platform

The CMSP platform is composed of a Nanopoli-100 polisher (Hangzhou Zhibang Nanotechnology Co., Ltd., Hangzhou, China), a lifting device, a fixture, an oscillating device, and an electric control device, as shown in Figure 1. The insert is fixed on the fixture with a screw. The fixture is connected to the oscillating device with the screws. The oscillating device is connected to the lifting device with the screws. The lifting device is mainly composed of a ball screw pair. The lifting device is adjusted to bring the insert edge in contact with the polishing pad and apply a certain polishing pressure. The polishing slurry is supplied by a peristaltic pump to drip onto the polishing pad at a certain flow rate. Meanwhile, the oscillating device is controlled by an electronic control device to swing the insert edge periodically within a certain angle range. The polishing slurry continuously flows through the insert edge with the rotation of the polishing pad. The oxidant in the polishing slurry reacts with the cutting edge to form a layer of oxide film. At the same time, the oxide film is removed by the mechanical action of the abrasive particles in the polishing slurry. The formation and removal of the film are alternated continuously. Therefore, the cutting edge preparation can be realized.

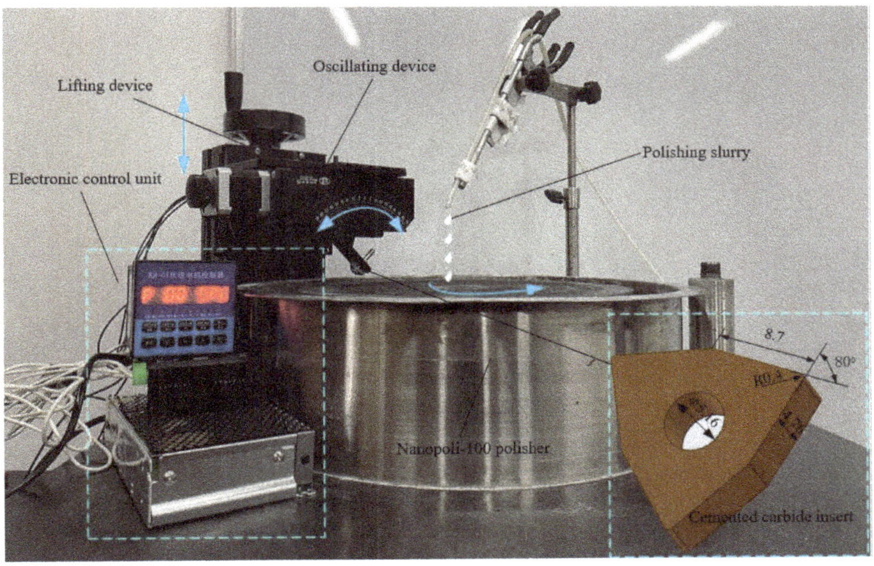

Figure 1. The CMSP platform.

3. Methods and Experiments

3.1. The Polishing Slurry Optimization Experiment

To achieve efficient and low-cost preparation of the carbide insert edge, the Taguchi method, combined with the grey relational analysis and fuzzy inference, is applied to optimize the CMSP polishing slurry for the cemented carbide inserts. The polishing object is a YG8 cemented carbide insert (Zhuzhou Cemented Carbide Cutting Tools Co., Ltd., Zhuzhou, China), as shown in Figure 2. Through the previous CMP experiments, the alumina abrasive particles are used as the abrasive particles, and the oxidant is H_2O_2 for the cemented carbide insert CMP, which can achieve high MRR and better surface quality [18]. In this study, the alumina abrasive particle with a particle size of 500 nm is chosen as the polishing abrasive particle. The oxidant is H_2O_2. The sodium dodecyl sulfate (SDS) is selected as the dispersant. Dilute sulfuric acid (H_2SO_4) and sodium hydroxide (NaOH) are used as pH regulators. The pH value of the slurry is measured using a digital PHS-3C Benchtop pH meter (Shaoxing Supo Instrument Co., Ltd., Shaoxing, China). Considering the dispersibility of the abrasive particles and the pH value of the polishing slurry on the MRR and surface roughness Ra, the abrasive particle concentration, oxidant concentration, dispersant concentration, and the pH value of the polishing slurry are selected as the main factors. The CMP orthogonal experiment for the insert is designed as a 4-factor 3-level Taguchi L9 orthogonal array, as shown in Table 1. The polishing experiment parameters are listed in Table 2. The polishing slurry is prepared according to Table 1. Then, the CMP experiment of the insert rake face is carried out on the Nanpoli-100 polisher. The polishing pad material is a porous polyurethane. A peristaltic pump supplies the polishing slurry with an adjustable flow rate. After the CMP of the inserts is finished, the inserts are cleaned in an ultrasonic cleaner with acetone for 5 min, then with anhydrous alcohol for 3 min, and finally with deionized water for 2 min. The stylus type surface roughness measuring instrument (JB-5C, Shanghai Taiming Optical Instrument Co., Ltd., Shanghai, China) with an accuracy of 0.001 μm is applied to measure the surface roughness Ra of the polished inserts three times, and the average value is regarded as the experimental data. Then, an electronic balance with an accuracy of 1/10,000 g is used to measure the weight of the inserts three times, and the average value is regarded as experimental data. The MRR of CMP for the cemented carbide insert is estimated as

$$MRR = \frac{\Delta M}{\rho \times S \times t}, \tag{1}$$

where ΔM is the difference in insert weight before and after polishing, ρ is the density of the insert, S is the contact area between the insert and the polishing pad, and t is the polishing time.

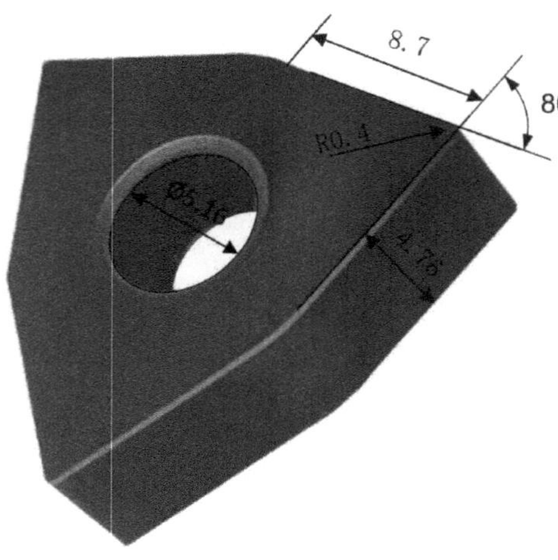

Figure 2. YG8 insert.

3.2. Cutting Edge Preparation Experiment

According to the optimal polishing slurry obtained in Section 3.1, the effect of the polishing plate's rotational speed, swing angle, and input frequency of the controller on the edge preparation process are investigated based on the CMSP platform. The polishing plate's rotational speed, swing angle, and input frequency of the controller are selected as the main factors. The orthogonal experiment of the edge preparation for the cemented carbide inserts is designed as a 3-factor 3-level Taguchi L9 orthogonal array, as shown in Table 3.

Table 1. The CMP orthogonal experiment for the insert.

No.	A, The Mass Concentration of Abrasive Particle (wt%)	B, The Mass Concentration of Oxidant (wt%)	C, The Mass Concentration of Dispersant (wt%)	D, pH
1	10	6	4	11
2	10	8	3	9.5
3	10	10	2	8
4	8	6	3	8
5	8	8	2	11
6	8	10	4	9.5
7	6	6	2	9.5
8	6	8	4	8
9	6	10	3	11

Table 2. CMP experiment parameters.

Parameter	Value
Polishing pressure (kPa)	94.99
Polishing plate's rotational speed (rpm)	70
Polishing time (min)	30
The flow rate of polishing slurry (mL/min)	6
Polishing slurry temperature (°C)	25

Table 3. The orthogonal experiment scheme of CMSP for the cemented carbide inserts cutting edge.

No.	A, Polishing Plate's Rotating Speed (rpm)	B, Swing Angle (°)	C, Input Frequency of the Controller (Hz)
1	30	6	7000
2	30	4	5000
3	30	2	3000
4	60	6	5000
5	60	4	3000
6	60	2	7000
7	90	6	3000
8	90	4	7000
9	90	2	5000

The YG8 cemented carbide inserts are selected for the preparation experiment. The preparation parameters are listed in Table 4.

Table 4. The parameters of CMSP for the cemented carbide insert cutting edge.

Parameters	Value
Polishing pressure (N)	5.5 (Linear cutting edge), 6 (Circular cutting edge)
Polishing slurry flow rate (mL/min)	4
Polishing slurry temperature (°C)	25
Preparation time (min)	5

The edge sharpness of the inserts after CMSP can be characterized by the cutting edge radius. Specifically, the cross-section intersection interception function in the VHX-2000 ultra-depth of field microscope system (KEYENCE, Tokyo, Japan) is used to construct a surface perpendicular to the edge, as shown in Figure 3. Then, the edge radius of the insert is obtained by sequentially selecting more than 30 points of the arc of the edge for the least squares fitting and calculation based on Matlab 2022. Taking the insert edge radius as the evaluation index, the edge preparation parameters are optimized by the Taguchi method. The effect of the polishing plate's rotational speed, swing angle, and input frequency of the controller on the edge preparation process are investigated by using ANOVA. The cutting edge morphology before and after preparation is observed by the VHX-2000 ultra-depth of field microscope and SEM.

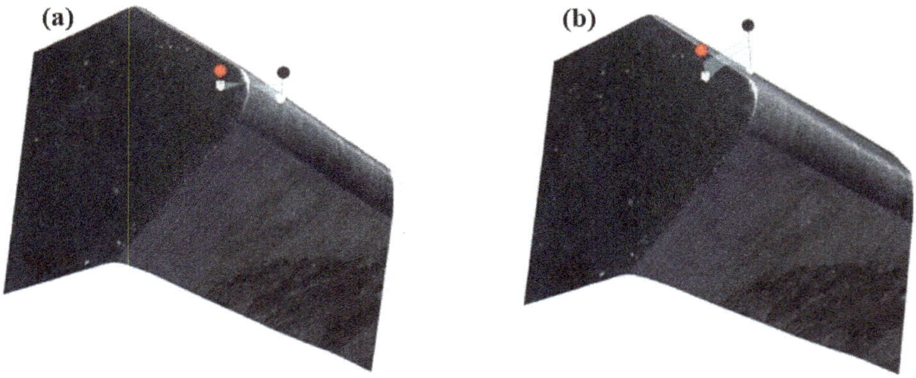

Figure 3. The surface perpendicular to the edge constructed in the VHX-2000 ultra depth of field microscope system: (**a**) circular edge, (**b**) linear edge.

4. Results and Discussions

4.1. The Optimization of the Polishing Slurry

4.1.1. The Calculation of the S/N Ratio of MRR and Surface Roughness Ra

The polishing slurry is prepared according to Table 1, and then the polishing experiments are carried out. The MRR and surface roughness Ra of the YG8 insert are obtained, as shown in Table 5.

Table 5. The experimental results of the MRR and surface roughness Ra of the insert.

No.	A, The Mass Concentration of Abrasive Particle (wt%)	B, The Mass Concentration of Oxidant (wt%)	C, The Mass Concentration of Dispersant (wt%)	D, pH	MRR (nm/min)	Surface Roughness Ra (nm)
1	10	6	4	11	66.85	47.33
2	10	8	3	9.5	74.81	47.5
3	10	10	2	8	79.58	27.83
4	8	6	3	8	39.79	31.5
5	8	8	2	11	55.71	27.33
6	8	10	4	9.5	57.3	35.5
7	6	6	2	9.5	41.38	40.17
8	6	8	4	8	50.93	41.33
9	6	10	3	11	30.24	46.5

In this study, it is expected that the MRR of the CMP for the cemented carbide insert should be as large as possible, and the surface roughness Ra of the polished cemented carbide insert should be as low as possible. Therefore, the larger the better case of the Taguchi method is applied to calculate the S/N ratio of the MRR, and the smaller-the-better case of the Taguchi method is used to calculate the S/N ratio of the surface roughness Ra, which can be expressed as follows:

$$\text{S/N ratio} = -10 \cdot \log\left[\frac{1}{n} \cdot \sum_{i=1}^{n} \frac{1}{{y_i}^2}\right], \quad (2)$$

$$\text{S/N ratio} = -10 \cdot \log\left[\frac{1}{n} \cdot \sum_{i=1}^{n} {y_i}^2\right], \tag{3}$$

where y_i is the experimental value; and n denotes the number of experiments. Calculation results are listed in Table 6.

Table 6. The S/N ratio of the MRR and surface roughness Ra.

No.	S/N Ratio of the MRR (dB)	S/N Ratio of the Surface Roughness Ra (dB)
1	36.5020	−33.5033
2	37.4791	−33.5339
3	38.0165	−28.8912
4	31.9959	−29.9662
5	34.9185	−28.7337
6	35.1631	−31.0046
7	32.3364	−32.0774
8	34.1402	−32.3259
9	29.6122	−33.3491

4.1.2. The Grey Rational Analysis

In the grey relational analysis method, when the range of the sequence is too large, the role of some factors is often ignored. When the target directions of each factor in the sequence are inconsistent, grey relational analysis may also result in wrong results. Firstly, it is necessary to perform data pre-processing for the entire sequence. The initial sequence is converted into a comparable sequence with the data pre-processing. The conversion is called grey relational generation. The initial sequence data are normalized in the range between zero and one. In this study, it is expected that the S/N ratio of the MRR and the surface roughness Ra should be as large as possible. The initial sequence data (S/N ratio of the MRR and the surface roughness Ra) are normalized with the "larger-the-better" characteristic of a linear data pre-processing method, which can be expressed as

$$x_i(k) = \frac{m_i(k) - \min m_i(k)}{\max m_i(k) - \min m_i(k)}, \tag{4}$$

where $x_i(k)$ is the value after data pre-processing, $m_i(k)$ denotes the initial sequence data, $\min m_i(k)$ denotes the minimum value of $m_i(k)$, $\max m_i(k)$ denotes the maximum value of $m_i(k)$. The normalized S/N ratios of MRR and surface roughness Ra are listed in Table 7.

Table 7. The normalized S/N ratios of MRR and surface roughness Ra.

No.	MRR	Surface Roughness Ra
1	0.8198	0.0064
2	0.9361	0
3	1	0.96702
4	0.2836	0.7432
5	0.6314	1
6	0.6605	0.5269
7	0.3241	0.3034
8	0.5388	0.2516
9	0	0.0385

Then, the grey rational coefficient $n_{ij}(k)$ is calculated based on the normalized S/N ratios of MRR and surface roughness Ra, which can be expressed as

$$n_{ij}(k) = \frac{\Delta_{\min} + \xi\Delta_{\max}}{\Delta_{ij}(k) + \xi\Delta_{\max}}, \tag{5}$$

where $\Delta_{ij}(k) = \|x_i(k) - x_j(k)\|$ denotes the deviation in absolute value between $x_i(k)$ and $x_j(k)$, $x_j(k) = 1(j = 1, 2, \ldots, 16)$ is the comparable sequence, ξ is the distinguishing coefficient, $\Delta_{\min} = \overset{\min}{\forall j} \in \overset{\min}{i \forall k}\|x_i(k) - x_j(k)\|$ is the minimum value of $\Delta_{ij}(k)$, $\Delta_{\max} = \overset{\max}{\forall j} \in \overset{\max}{i \forall k}\|x_i(k) - x_j(k)\|$ is the maximum value of $\Delta_{ij}(k)$. The grey rational coefficients of the MRR and Ra are listed in Table 8.

Table 8. The grey relational coefficients of the MRR and surface roughness Ra.

No.	MRR	Surface Roughness Ra
1	0.7351	0.3348
2	0.8866	0.3333
3	1.0000	0.9384
4	0.4111	0.6607
5	0.5756	1.0000
6	0.5956	0.5138
7	0.4252	0.4179
8	0.5202	0.4005
9	0.3333	0.3421

4.1.3. Fuzzy Inference

The inference system based on fuzzy rules consists of three basic units: fuzzier, fuzzy inference engine, and defuzzifier. The input vector of the gray relational coefficients for MRR and the surface roughness Ra is first fuzzified by the fuzzifier using the membership functions. Then, the fuzzy inference is performed by the fuzzy inference engine using the fuzzy rules to generate the fuzzy values. Finally, the fuzzy values are transformed to the multi-performance characteristics index (MPCI). In this study, the gray relational coefficients of MRR and the surface roughness Ra are regarded as the input variable for the fuzzy inference system (FIS), respectively. The MPCI value is regarded as the output variable. Based on MATLAB applications, the FIS architecture is established. The triangular membership function is chosen for the input variable. There are three membership functions, named small (S), middle (M), and large (L), respectively. The triangular membership function is selected for the output variable MPCI, and there are five membership functions, named very small (VS), small (S), middle (M), large (L), and very large (VL), respectively. Therefore, nine fuzzy rules are generated in the rule editor of MATLAB, that is,

$$
\begin{aligned}
&\text{Rule 1}: \text{ if m is S and r is S then y is VS;}\\
&\text{Rule 2}: \text{ if m is S and r is M then y is S;}\\
&\text{Rule 3}: \text{ if m is S and r is L then y is M;}\\
&\text{Rule 4}: \text{ if m is M and r is S then y is S;}\\
&\text{Rule 5}: \text{ if m is M and r is M then y is M;}\\
&\text{Rule 6}: \text{ if m is M and r is L then y is L;}\\
&\text{Rule 7}: \text{ if m is L and r is S then y is M;}\\
&\text{Rule 8}: \text{ if m is L and r is M then y is L;}\\
&\text{Rule 9}: \text{ if m is L and r is L then y is VL;}
\end{aligned}
$$

where y is the output, m and r are the gray relational coefficients inputs for MRR and surface roughness Ra, respectively. Then, the MPCI of each parameter combination is obtained based on the FIS of Matlab, as listed in Table 9.

Table 9. The MPCI of each parameter combination.

No.	1	2	3	4	5	6	7	8	9
MPCI	0.524	0.580	0.864	0.532	0.754	0.559	0.443	0.454	0.392

4.1.4. The Optimization of Polishing Slurry

To obtain the overall optimal parameter combination of polishing slurry, combining Table 5 with Table 9, the MPCI response for each factor at their corresponding levels is obtained, as shown in Figure 4.

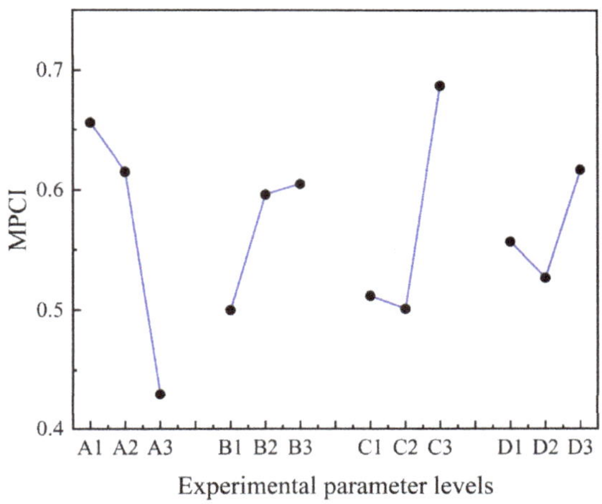

Figure 4. The MPCI values for each factor at their corresponding levels.

In Figure 4, it can be seen that A1B3C3D3 is the best parameter combination of polishing slurry for cemented carbide inserts. Namely, the best parameter combination of polishing slurry for cemented carbide inserts is with the mass concentration of the abrasive particle of 10 wt%, the mass concentration of the oxidant of 10 wt%, the mass concentration of the dispersant of 2 wt%, and a pH of 8. As the concentration of abrasive particles increases, the MPCI also increases. The reason is that as the concentration of abrasive particles increases, the number of effective abrasive particles involved in material removal increases, leading to an increase in MRR and a decrease in surface roughness Ra. Furthermore, as the concentration of the oxidant increases, the MPCI also increases. The reason is that as the concentration of oxidant increases, the chemical reaction between the surface of the cemented carbide insert and the oxidant is promoted, and MRR increases. The abrasive particles tend to aggregate and settle in the polishing slurry, and have an impact on MRR and surface quality. The abrasive particles can be dispersed by the dispersant without affecting the thermal conductivity of the polishing slurry, thereby ensuring the high dispersion stability of the polishing slurry. As the concentration of dispersant increases, MPCI first decreases and then slightly increases. The reason is that when the dispersant dosage is too high, the molecular chains of unabsorbed dispersant in the solvent tend to entangle with each other due to the supersaturated adsorption on the surface of the alumina particles, leading to depletion flocculation and agglomeration of abrasive particles, thereby affecting MRR and surface quality [18]. The pH of the suspension has a significant impact on the dispersion of the polishing slurry and the chemical reaction between the insert surface and the oxidant [19], thereby affecting the MRR and surface quality. When the pH value is eight, the MPCI reaches the maximum value. It indicates that when the pH value is eight, the dispersion of the polishing slurry and the chemical reaction between the insert surface and the oxidant may be in a good state. Compared with the initial parameter combination A3B3C2D1, MRR is increased by 163%, and the surface roughness Ra is reduced by 40%. Then, the insert rake surface morphology before and after polishing with the best parameter combination of polishing slurry is observed by SEM, as shown in Figure 5.

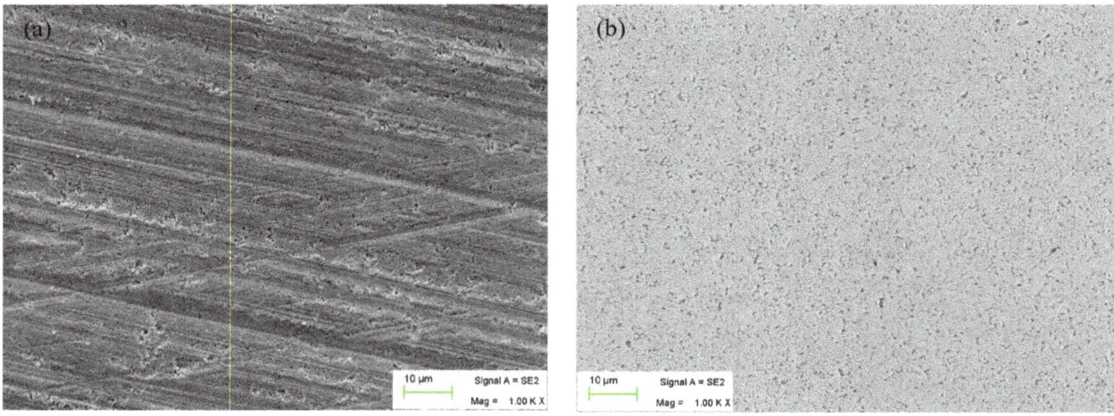

Figure 5. The insert rake surface morphology under SEM: (**a**) before polishing, (**b**) after polishing.

In Figure 5, it can be seen that there is a lot of scratches on the insert rake surface before polishing. However, after the insert rake surface is polished with the best parameter combination of polishing slurry, the scratches are completely removed, and the rake surface is smooth. So the best parameter combination of polishing slurry is effective and can obtain a smooth surface.

4.2. Comprehensive Analysis of CMSP for the Cemented Carbide Insert Edge

According to Table 3, the orthogonal experiment of CMSP for the cemented carbide insert cutting edge is carried out. Then, the edge radius of the insert is obtained by the least squares fitting and calculation based on Matlab, which are listed in Table 10.

Table 10. The experimental results of CMSP for the cemented carbide insert cutting edge.

No.	A, Polishing Plate's Rotational Speed (rpm)	B, Swing Angle (°)	C, Input Frequency of the Controller (Hz)	Linear Edge Radius (μm)	Circular Edge Radius (μm)
1	30	6	7000	46.895	70.619
2	30	4	5000	34.508	54.862
3	30	2	3000	33.011	49.780
4	60	6	5000	71.259	80.616
5	60	4	3000	41.822	57.024
6	60	2	7000	60.273	74.764
7	90	6	3000	67.279	78.155
8	90	4	7000	46.938	73.451
9	90	2	5000	59.435	74.113

It is desirable that the edge radius should be as large as possible in the same preparation time. So the larger the better case of the Taguchi method is applied to calculate the S/N ratio of the linear edge radius and circular edge radius, which are listed in Table 11. Figure 6 shows the average response of S/N for each factor at their corresponding levels.

Table 11. The S/N ratio of the linear edge radius and circular edge radius.

No.	A, Polishing Plate's Rotational Speed (rpm)	B, Swing Angle (°)	C, Input Frequency of the Controller (Hz)	SNR of the Linear Edge Radius (dB)	SNR of the Circular Edge Radius (dB)
1	30	6	7000	33.423	36.978
2	30	4	5000	30.758	34.785
3	30	2	3000	30.373	33.941
4	60	6	5000	37.057	38.128
5	60	4	3000	32.428	35.121
6	60	2	7000	35.602	37.474
7	90	6	3000	36.558	37.859
8	90	4	7000	33.430	37.320
9	90	2	5000	35.481	37.398

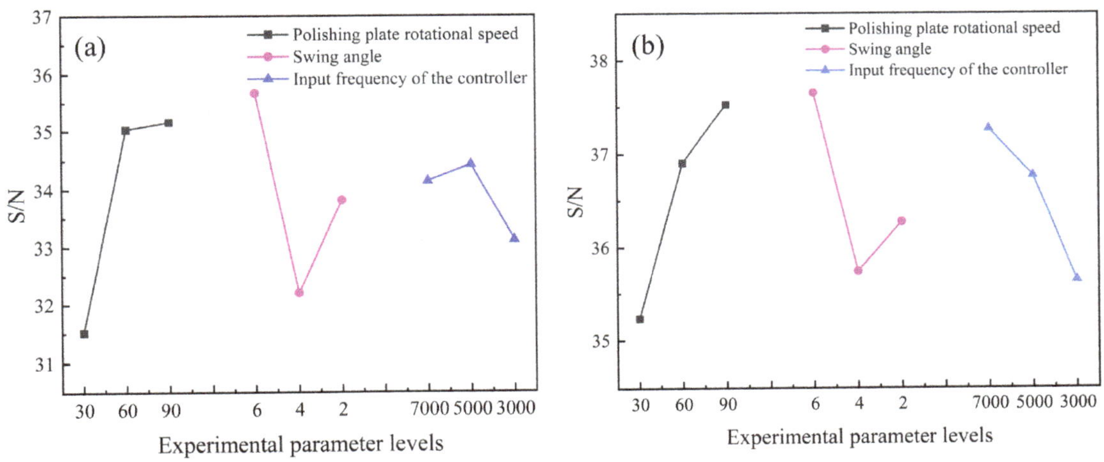

Figure 6. The average response of S/N for each factor at their corresponding levels: (**a**) linear edge, (**b**) circular edge.

It is noted in Figure 6 that the best combination of CMSP parameters for the linear edge and circular edge is determined to be A3B1C2 and A3B1C1, respectively. Namely, the best CMSP process parameter combination for the linear edge is with the polishing plate's rotational speed of 90 rpm, the swing angle of 6°, and the input frequency of the controller of 5000 Hz, and the best CMSP process parameter combination for the circular edge is with the polishing plate's rotational speed of 90 rpm, the swing angle of 6°, and the input frequency of the controller of 7000 Hz.

To explore the effect of the polishing plate's rotational speed, swing angle, and input frequency of the controller on the cutting edge radius, based on Table 10, ANOVA results of CMSP for the insert linear and circular edge radius are shown in Tables 12 and 13, respectively.

Table 12. ANOVA results of CMSP for the insert linear edge radius.

Source	Degree of Freedom	Sum of Squares	Mean Square	F Value	*p* Value	Contribution (%)
A	2	775.89	387.945	21.784	0.044	47.91
B	2	644.679	322.34	18.1	0.052	39.42
C	2	88.897	44.448	2.496	0.286	3.45
Error	2	35.617	17.809			9.22
Total	8	1545.083				100

Significant at 95% confidence interval.

Table 13. ANOVA results of CMSP for the insert circular edge radius.

Source	Degree of Freedom	Sum of Squares	Mean Square	F Value	*p* Value	Contribution (%)
A	2	455.878	227.939	73.068	0.014	44.66
B	2	340.29	170.145	54.542	0.018	33.18
C	2	204.399	102.199	32.761	0.03	19.68
Error	2	6.239	3.12			2.48
Total	8	1006.806				100

Significant at 95% confidence interval.

From Tables 12 and 13, it can be seen that the polishing plate's rotational speed plays a predominant role in the linear edge and circular edge preparation, followed by the swing angle and the input frequency of the controller.

The cutting edge is prepared with the best combination of CMSP parameters for the linear edge and circular edge of the insert. Then, the cutting edge surface morphology before and after preparation is observed by the VHX-2000 ultra depth of field microscope and SEM, as shown in Figures 7 and 8.

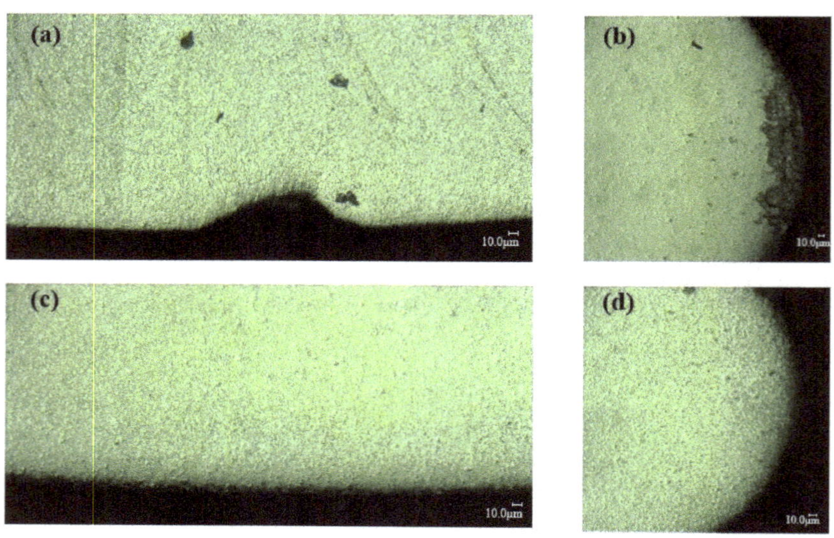

Figure 7. The morphology of the cutting edge under the VHX-2000 ultra depth of field microscope: (**a**) linear edge before the preparation, (**b**) circular edge before the preparation, (**c**) linear edge after the preparation, (**d**) circular edge after the preparation.

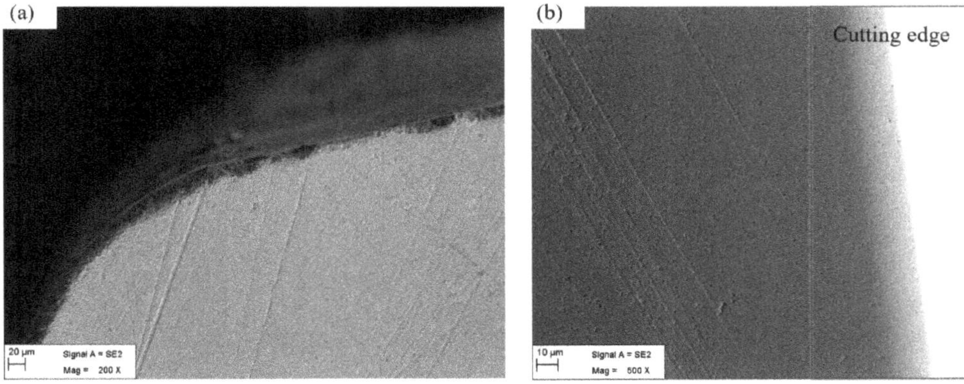

Figure 8. The morphology of the cutting edge under SEM: (**a**) cutting edge before the preparation, (**b**) cutting edge after the preparation.

In Figures 7 and 8, it can be seen that the initial cutting edge has some defects in the edge region such as burrs, poor surface finish, micro-breakages, and irregularities along the edge. After the cutting edge is prepared with the best combination of CMSP parameters, the defects are completely removed, and the cutting edge is smooth. So CMSP can improve the surface quality of the cutting edge. Therefore, in the actual cutting process, it can effectively reduce tool wear, improve the machining process stability, and reduce the machined surface quality.

5. Conclusions

In this work, the CMSP device for the cemented carbide insert cutting edges is constructed. Then, the polishing slurry of CMSP for cemented carbide insert cutting edge is optimized using the Taguchi method coupled with the grey relational analysis and fuzzy inference. Finally, the orthogonal experiments, the Taguchi method, and ANOVA are used to investigate the effect of the polishing plate's rotational speed, swing angle, and input frequency of the controller on the edge preparation process, and the parameters are optimized. Thus, the following conclusions can be drawn:

1. A1B3C3D3 is the optimal parameter combination of polishing slurry for cemented carbide inserts. Namely, the best parameter combination of polishing slurry for cemented carbide inserts is with the mass concentration of the abrasive particle of 10 wt%, the mass concentration of the oxidant of 10 wt%, the mass concentration of the dispersant of 2 wt%, and the pH of 8. Compared with the initial parameter combination, A3B3C2D1, the MRR is increased by 163%, and the surface roughness Ra is reduced by 40%.
2. The best combination of CMSP parameters for the linear edge and circular edge is determined to be A3B1C2 and A3B1C1, respectively. Namely, the best CMSP process parameter combination for the linear edge is with the polishing plate's rotational speed of 90 rpm, the swing angle of 6°, and the input frequency of the controller of 5000 Hz, and the optimum CMSP process parameter combination for the circular edge is with the polishing plate's rotational speed of 90 rpm, the swing angle of 6°, and the input frequency of the controller of 7000 Hz.
3. The polishing plate's rotational speed plays a predominant role in linear edge and circular edge preparation, followed by the swing angle and the input frequency of the controller.
4. The CMSP method can not only achieve cutting edge preparation, but also improve the surface quality of the cemented carbide insert cutting edge. In the future, extensive

research is required to elucidate the material removal mechanism of CMSP for the cemented carbide insert cutting edge.

Author Contributions: Conceptualization, C.Q.; methodology, C.Q.; writing—original draft, C.Q. and J.P.; writing—review and editing, C.Q., Z.H., M.M. and S.J.; investigation, C.Q., J.P., L.G. and W.C.; formal analysis, C.Q. and W.C.; data curation, C.Q., C.Z., X.C. and W.C.; supervision, Z.H., M.M. and S.J.; software, C.Q. and W.C. All authors have read and agreed to the published version of the manuscript.

Funding: This work is supported by the Hunan Provincial Natural Science Foundation of China (No. 2022JJ50134, 2022JJ50133, 2021JJ20009, 2021JJ50125) and the program of Xiangtan Science and Technology Bureau of China (No. CG-YB20221040), which the authors greatly appreciate.

Data Availability Statement: The authors confirm that the data supporting the findings of this study are available within the article.

Conflicts of Interest: The authors declare no conflict of interest.

References

1. Schneider, F.; Das, J.; Kirsch, B.; Linke, B.; Aurich, J.C. Sustainability in ultra precision and micro machining: A review. *Int. J. Precis. Eng. Manuf.—Green Technol.* **2019**, *6*, 601–610. [CrossRef]
2. Qu, S.; Yao, P.; Gong, Y.; Chu, D.; Yang, Y.; Li, C.; Wang, Z.; Zhang, X.; Hou, Y. Environmentally friendly grinding of C/SiCs using carbon nanofluid minimum quantity lubrication technology. *J. Clean. Prod.* **2022**, *36*, 132898. [CrossRef]
3. Li, C.; Piao, Y.; Meng, B.; Hu, Y.; Li, L.; Zhang, F. Phase transition and plastic deformation mechanisms induced by self-rotating grinding of GaN single crystals. *Int. J. Mach. Tools Manuf.* **2022**, *172*, 103827. [CrossRef]
4. Denkena, B.; Köhler, J.; Ventura, C.E.H. Customized cutting edge preparation by means of grinding. *Precis. Eng.* **2013**, *37*, 590–598. [CrossRef]
5. Denkena, B.; de Leon, L.; Bassett, E.; Rehe, M. Cutting edge preparation by means of abrasive brushing. *Key Eng. Mater.* **2010**, *438*, 1–7. [CrossRef]
6. Yussefian, N.Z.; Koshy, P.; Buchholz, S.; Klocke, F. Electro-erosion edge honing of cutting tools. *CIRP Ann.* **2010**, *59*, 215–218. [CrossRef]
7. Zhang, S.; Zou, B.; Liu, Y.; Wang, Y.; Huang, C.; Liu, Z. Edge passivation and quality of carbide cutting inserts treated by wet micro-abrasive blasting. *Int. J. Adv. Manuf. Technol.* **2018**, *96*, 2307–2318. [CrossRef]
8. Wang, W.; Biermann, D.; Aßmuth, R.; Arif, A.F.M.; Veldhuis, S.C. Effects on tool performance of cutting edge prepared by pressurized air wet abrasive jet machining (PAWAJM). *J. Mater. Process. Technol.* **2020**, *277*, 116456. [CrossRef]
9. Wang, W.; Saifullah, M.K.; Aßmuth, R.; Biermann, D.; Arif, A.F.M.; Veldhuis, S.C. Effect of edge preparation technologies on cutting edge properties and tool performance. *Int. J. Adv. Manuf. Technol.* **2020**, *106*, 1823–1838. [CrossRef]
10. Zimmermann, M.; Kirsch, B.; Kang, Y.; Herrmann, T.; Aurich, J.C. Influence of the laser parameters on the cutting edge preparation and the performance of cemented carbide indexable inserts. *J. Manuf. Process.* **2020**, *58*, 845–856. [CrossRef]
11. Lyu, B.; Ke, M.; Fu, L.; Duan, S.; Shao, Q.; Zhou, Y.; Yuan, J. Experimental study on the brush tool–assisted shear-thickening polishing of cemented carbide insert. *Int. J. Adv. Manuf. Technol.* **2021**, *115*, 2491–2504. [CrossRef]
12. Shao, L.; Zhou, Y.; Fang, W.; Wang, J.; Wang, X.; Deng, Q.; Lyu, B. Preparation of cemented carbide insert cutting edge by flexible fiber-assisted shear thickening polishing method. *Micromachines* **2022**, *13*, 1631. [CrossRef] [PubMed]
13. Ali, I.; Roy, S.R.; Shinn, G. Chemical-mechanical polishing of interlayer dielectric: A review. *Solid State Technol.* **1994**, *37*, 63–68.
14. Malik, F.; Hasan, M. Manufacturability of the CMP process. *Thin Solid Film.* **1995**, *270*, 612–615. [CrossRef]
15. Hu, Z.; Qin, C.; Chen, X.; Tang, A.; Fang, T.; Yang, Z.; Mao, M. Chemical–mechanical polishing of cemented carbide insert surface for extended tool life in turning of GH4169 nickel-based superalloy. *Int. J. Precis. Eng. Manuf.* **2020**, *21*, 1421–1435. [CrossRef]
16. Hu, Z.; Qin, C.; Chen, Z.C.; Yang, Z.; Fang, T.; Mao, M. Experimental study of chemical mechanical polishing of the final surfaces of cemented carbide inserts for effective cutting austenitic stainless steel. *Int. J. Adv. Manuf. Technol.* **2018**, *95*, 4129–4140. [CrossRef]
17. Qin, C.; Hu, Z.; Tang, A.; Yang, Z.; Luo, S. An efficient material removal rate prediction model for cemented carbide inserts chemical mechanical polishing. *Wear* **2020**, *452*, 203293. [CrossRef]
18. Seo, J.; Paik, U. Preparation and characterization of slurry for chemical mechanical planarization (CMP). In *Advances in Chemical Mechanical Planarization;* CMP: Augusta, ME, USA, 2016; pp. 273–298.
19. Singh, B.P.; Menchavez, R.; Takai, C.; Fuji, M.; Takahashi, M. Stability of dispersions of colloidal alumina particles in aqueous suspensions. *J. Colloid Interface Sci.* **2005**, *291*, 181–186. [CrossRef] [PubMed]

Disclaimer/Publisher's Note: The statements, opinions and data contained in all publications are solely those of the individual author(s) and contributor(s) and not of MDPI and/or the editor(s). MDPI and/or the editor(s) disclaim responsibility for any injury to people or property resulting from any ideas, methods, instructions or products referred to in the content.

micromachines

Article

Investigation of Surface Integrity Induced by Ultra-Precision Grinding and Scratching of Glassy Carbon

Kirk Jahnel [1,*], Robert Michels [1], Dennis Patrick Wilhelm [1], Tim Grunwald [1] and Thomas Bergs [1,2]

[1] Department of Fine Machining and Optics, Fraunhofer Institute for Production Technology IPT, 52074 Aachen, Germany; robert.michels@ipt.fraunhofer.de (R.M.); dennis.patrick.wilhelm@ipt.fraunhofer.de (D.P.W.); tim.grunwald@ipt.fraunhofer.de (T.G.); t.bergs@wzl.rwth-aachen.de (T.B.)

[2] Chair for Manufacturing Technology, Tool Machine Laboratory (WZL), RWTH Aachen University, 52074 Aachen, Germany

* Correspondence: kirk.jahnel@ipt.fraunhofer.de; Tel.: +49-241-8904-127

Abstract: Glassy carbon provides material characteristics that make it a promising candidate for use as a mould material in precision glass moulding. However, to effectively utilize glassy carbon, a thorough investigation into the machining of high-precision optical surfaces is necessary, which has not been thoroughly investigated. This research analyses the process of material removal and its resulting surface integrity through the use of nano-scratching and ultra-precision grinding. The nano-scratching process begins with ductile plastic deformation, then progresses with funnel-shaped breakouts in the contact zone, and finally concludes with brittle conchoidal breakouts when the cutting depth is increased. The influence of process factors and tool-related parameters resulting from grinding has discernible impacts on the ultimate surface roughness and topography. Enhancing the cutting speed during cross-axis kinematic grinding results in improved surface roughness. Increasing the size of diamond grains and feed rates leads to an increase in surface roughness. An achievable surface roughness of Ra < 5 nm together with ductile-regime grinding behaviour meet optical standards, which makes ultra-precision grinding a suitable process for optical surface generation.

Keywords: ultra-precision grinding; nano-scratch; surface integrity; ductile grinding; brittle material; glassy carbon

1. Introduction

In contemporary times, a substantial quantity of sensors is integrated into widely used commodities such as smartphones and automobiles. A significant portion of these sensors rely on MEMS due to their compact dimensions and efficient energy usage, which enable a wide range of sensory functions. The integration of borosilicate glass optical components, such as protective coverings and beam guides, with MEMS, is facilitating the development of compact optoelectronic applications [1]. These systems are made possible because of the thermal expansion properties of borosilicate glass types, which are specifically tailored to be compatible with silicon. As a result, the risk of stress formation during the joining process can be mitigated. Modern applications necessitate the use of intricate glass components with advanced optical properties. Precision glass moulding is a well-established technological process utilized for the effective production of optics. This process involves the shaping of a glass preform using moulds that possess a high degree of precision [2,3]. However, the moulding temperature of borosilicate glass (>700 °C) is higher compared to other types of glass (500–600 °C). As a result, the existing mould coatings used for precision glass moulding have a shorter lifespan when subjected to the temperature required for moulding borosilicate glass [4].

One potential option is the utilization of glassy carbon, a high-temperature material that possesses a combination of features resembling both graphite and diamond. Notably, glassy carbon exhibits chemical and physical characteristics that eliminate the need for a coating [5,6]. One of the primary difficulties encountered pertains to the production of moulds utilizing glassy carbon material. The process of ultra-precision grinding, which has previously been established in the manufacturing industry for producing moulds with optical surfaces, is deemed to be a suitable method [7]. The theory of ductile-regime grinding is a frequently used approach that facilitates the achievement of defect-free surfaces through a regulated and deterministic process [8,9].

Nevertheless, the precise nature of the cutting behaviour and the subsequent formation of micro flaws during the machining of glassy carbon remains uncertain. Hence, the primary objective of this study is to comprehensively comprehend the underlying factors contributing to the occurrence of defects in the process of ultra-precision grinding of glassy carbon. This research specifically examines the cutting mechanisms involving diamond cutting edges and their impact on the surface integrity of glassy carbon.

2. Materials and Methods

Glassy carbon is classified as an inorganic non-metallic substance. The phrase "non-organic materials" pertains to substances that possess a structure without organic components, wherein primary heteropolar or covalent bonds are present. The interior structure comprises lattice or mesh structures, as seen in Figure 1 [10].

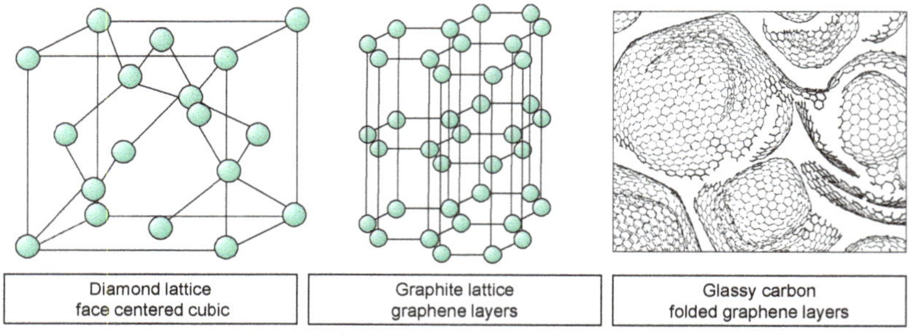

Figure 1. Visualization of the lattice structure of diamond, graphite, and glassy carbon.

Carbon is found in two stable crystalline forms; namely, diamond and graphite. In the process of diamond modification, the carbon atoms arrange themselves in a face-centred cubic (FCC) lattice. The structure of diamond is one of the contributing factors to its exceptional hardness, among other characteristics [10]. Graphite is the term used to describe the most advanced and stable alteration of carbon material. Graphite exhibits a distinctive feature of having consistently organized hexagonal layer lattices, which are also referred to as graphene. Strong covalent connections exist within these levels. The interlayer bonding between the layers is characterized by a relatively low strength. The structural characteristics of the material give rise to its direction-dependent features. In contrast to the conventional layer arrangement observed in graphite, there exist diverse microstructures wherein the graphene layers exhibit a state of disorder. The materials in question are referred to as paracrystalline carbon materials [10,11].

Within these paracrystalline formations, the distribution of graphene fragments often follows a normal distribution with regard to their shape and size. The presence of flaws in the microstructure leads to the folding and interweaving of individual graphene layers in a three-dimensional manner, ultimately leading to the formation of fullerene-like structures. According to the works of Harris [12] and Jurkiewicz [13], it is evident that glassy carbon is characterized by its paracrystalline nature and closed-porous microstructure, which is mostly composed of interlaced and folded hexagonal layers. The microstructure exhibits variations contingent upon the specific manufacturing method employed, hence resulting in corresponding variations within a defined range of material attributes [14]. Currently, there is not a universally acknowledged structural framework for glassy carbon [15].

Glassy carbon has a high degree of brittleness often characterized by flat surfaces in its untreated state. According to Sharma [14], glassy carbon exhibits impermeability to a wide range of gases and liquids and demonstrates chemical inertness. Glassy carbon is a kind of carbon that is produced from the pyrolysis of polymers, and it does not form graphite [12]. During this procedure, the polymer chains undergo fragmentation due to the application of high temperatures, resulting in the creation of novel molecules involving carbon atoms exclusively. The formation of different carbon modifications or chars during pyrolysis is influenced by factors such as the polymer employed, the atmospheric conditions, and the temperature. In practical applications, the pyrolysis of glassy carbon involves the use of strongly crosslinked aromatic polymers, such as phenolic and furan resins [11].

Glassy carbon exhibits excellent corrosion resistance, impermeability to gases and liquids, and high thermal stability, withstanding temperatures of up to 550 °C in ambient air and up to 3000 °C when shielded in a protective atmosphere. The fracture toughness of the material is measured to be 20 $N/mm^{3/2}$, while the modulus of elasticity is determined to be 35 kN/mm^2. Glassy carbon is utilized as electrode material in laboratory technology, in medicinal technology, and in high-temperature applications in mechanical engineering due to its inherent qualities [11].

Table 1 presents a comprehensive compilation of material parameters pertaining to glassy carbon, with a comparative analysis of optical glass BK7. The machining characteristics of BK7 optical glass have been thoroughly studied and serve as a foundation for the examination of glassy carbon in this research. Despite its name, glassy carbon displays distinct material characteristics from glass. However, a comparison between the two is still valuable because of their shared brittle material behaviour.

Table 1. Material properties of glassy carbon and BK7 [16,17].

Material Property	Glassy Carbon Sigradur G	Optical Glass BK 7
Characteristic temperature	>3000 °C (decomposition)	557 °C (transition)
Density	1.42 g/cm^3	2.51 g/cm^3
Hardness	2.26 GPa	5.80 GPa
Young's modulus	35 GPa	82 GPa
Thermal conductivity	6.30 W/(mK)	1.11 W/(mK)
Coefficient of thermal expansion (20–200 °C)	2.6×10^{-6} /K	8.3×10^{-6} /K

Indentation, Nano-Scratch, and Grinding Research Method

In the process of developing and manufacturing mechanical parts and components, mechanical parameters such as yield stress have significant importance. The mechanical properties of glass, in particular, exhibit variations based on their chemical composition and production techniques [18]. Conventional mechanical testing techniques, such as the tensile test [19], are intricate and costly, rendering them unsuitable for evaluating miniature components due to the challenges associated with specimen preparation. The methods employed in this particular case involve indentation and scratch testing. Several research studies have been undertaken utilizing the indentation method to evaluate the elastic-plastic characteristics of materials by analysing their indentation behaviour [20–22]. The relationship between plastic stress and indentation hardness at different strains is a valuable tool for establishing a connection and determining the yield stress and other plastic properties of metals.

Additional experimental studies utilizing modified hardness indentation tests on various optical glasses demonstrate consistent fracture propagation behaviour over a wide range of glass compositions. During the loading process, a hydrostatic compressive stress field is generated beneath the body of the pyramidal indentation [23]. The evolution of the material is characterized by both an increase in the density of the glass and the construction of a wall composed of plasticized material along the surfaces of the pyramid at a macroscopic level. Stress-relief fractures characterized by a rapid propagation velocity and perpendicular direction to the glass surface along the indentation diagonal manifest abruptly alone during the process of stress release [24]. When the glass surface experiences complete relief, the formation of fractures occurs at a relatively moderate pace. The application of pressure by the indentation body causes the glass to undergo compression, resulting in the development of radial tensile stresses. They have the responsibility for the continued dissemination of the fractures in a conchoidal form.

With the method of scratch testing, similar mechanisms of the above-mentioned also occur when a glass surface is subjected to stress during scratching. The scratch test method should be employed to examine the plastic characteristics of materials [25]. Scratching is a well-known empirical method for determining the hardness according to the Mohs scale [26,27]. Because it has a negative effect on the service life of glasses [28–30], if the scratch normal force does not exceed a certain level, a crack-free, plastically deformed scratch trace can be produced. Due to the stresses induced in the vicinity of the scratch by the compaction zone, such plastic scratches are highly unstable and tend to form cracks. The cracks spread into the existing plastic part of the scratch and form conchoidal side cracks.

The investigation of brittle behaviour is commonly conducted using various materials by indentation and nano-scratch techniques; therefore, Bifano's model is frequently utilized as a reference for determining the critical depth of cut in ductile-regime machining of brittle materials [8]. Nevertheless, the model solely relies on material characteristics and does not include factors like machining conditions, tool shape, and size [31]. Adjusting the process settings to decrease the chip thickness leads to improved surface quality, as it approaches or decreases below the critical chip thickness for ductile grinding [32–35]. The alteration of the grain size of the grinding tool also affects the resulting chip thickness. Smaller diamond grains lead to improved surface roughness and might potentially induce complete plastic material behaviour [35]. However, further examinations utilizing nano-scratching techniques in gallium and silicon demonstrate that the essential cutting depths varied depending on whether conical or Berkovich tips are employed [36]. Modifying the lubrication quantity can also result in corresponding effects on the critical chip thickness [37], unrelated to the material properties. The anisotropy of the material significantly affects the damage evolution and material removal behaviours during the machining process [38].

To examine the aforementioned processes in glassy carbon, a series of traditional Vickers hardness tests and nano-scratch examinations were conducted on prepared polished samples. The schematic technique represented in Figure 2 enables the investigation of whether glassy carbon has comparable brittle material behaviour. The scratch examinations were conducted at several cutting speeds to yield results for the subsequent grinding research. Furthermore, a scratch trace was subjected to calotte grinding in order to observe the presence of the separation phenome in the subsurface.

The motion of the cutting edge in the scratch tests may be described as a helical route. In this setup, the feed movement with the workpiece velocity v_w is achieved by utilizing the z-axis (movement direction: $-z$), on which the glassy carbon sample is securely affixed to a sample holder. The helix undergoes a circular motion due to the continuous rotation of the spindle. The helix's central axis is curved towards the sample by a concurrent semicircular motion in the negative y direction. If there is enough contact with the sample surface, the cumulative movement of all axes will create scratches on the surface of the sample. These scratches will vary in length and depth. Please refer to Figure 2 in the upper right corner for a visual representation. As a result of the continuous feed along the z-axis, all scratches have an equal path distance a_p. The experiments were conducted using a typical UP-grinding machine; namely, the Nanotech 350FG model (Moore Nanotechnology Systems, LLC, Swanzey, NH, USA). In order to obtain the necessary cutting speeds v_c, a customized tool holder was devised and affixed to the machine's C-axis spindle. Performing scratch testing on the same grinding machine can enhance the capacity to replicate the grinding process behaviour, as it also replicates factors such as machine vibrations. The uniform machine environment should further enhance the transferability to other experiments in this publication. A sharply cut diamond from a turning tool was used as the abrasive grain. A diamond with an included angle of 90° and a rake angle of −35° was used to simulate the geometry of a diamond grain.

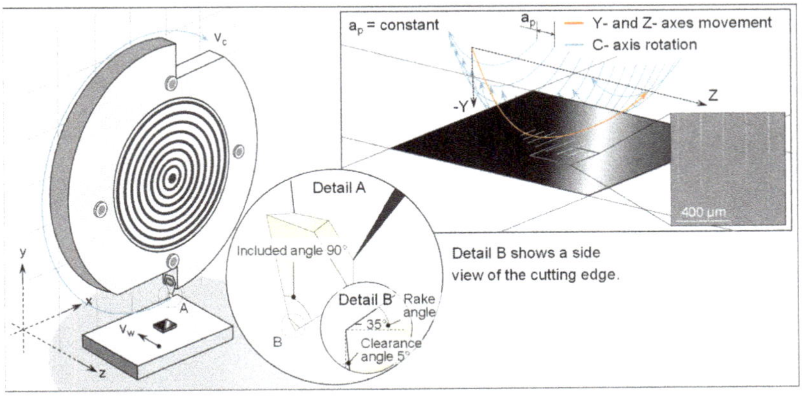

Figure 2. Visualization of the scratch test setup and kinematics.

A restriction in the investigation of material removal mechanisms by means of indentation or scratch tests is that only the singular intervention of a grain can be investigated. In the grinding process, however, there are interactions and overlaps of engagement paths of differently shaped diamond grains, which can thus constantly change the removal mechanisms. Consequently, a series of comprehensive grinding technology experiments was conducted. The experiments were conducted using a 5-axis ultra-precision grinding machine. The machine is situated inside a regulated setting characterized by a consistent temperature (T = 21 ± 0.1 °C) and relative humidity (40 ± 5%). The schematic representation of the experimental system is illustrated in Figure 3. Liquid isoparaffin is employed as a cooling agent in the context of minimal amount lubrication. The use of isoparaffin has become widely accepted as the prevailing method in ultra-precision machining [39].

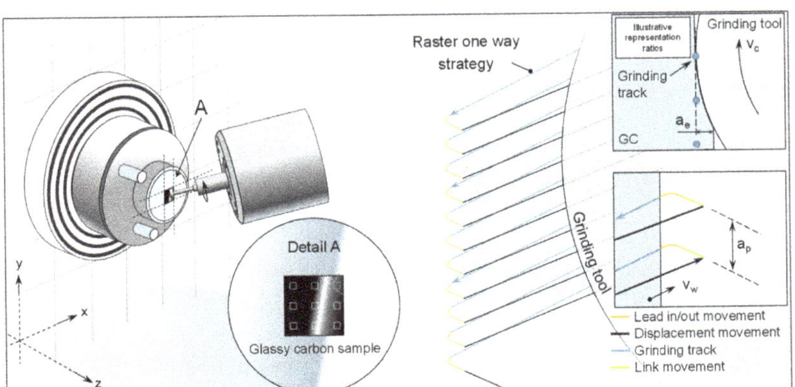

Figure 3. Visualization of grinding test setup and strategy.

The glassy carbon Sigradur G utilized in this study was procured from HTW Hochtemperatur Werkstoffe GmbH (Thierhaupten, Germany) [16]. The specimens exhibit a square shape, measuring 10 mm × 10 mm in length and width, with a thickness of 3 mm. In order to provide equitable testing conditions and reduce any potential surface imperfections, the specimens undergo a preliminary grinding and polishing process. The resultant initial surface has a consistently polished appearance, characterized by roughness values Sq that are less than 2 nm.

Prior to each usage, the grinding wheels undergo dressing to achieve a tip radius of 400 µm. In order to avoid the grinding tools from becoming unclamped from the grinding spindle during the dressing process, they are dressed on a dedicated dressing spindle that is mounted within the machine. This offers the benefit of utilizing high-precision machine axes to achieve precise dressing. The dressing procedure consists of two stages. Prior to the application of the dressing, any irregularities are eliminated and the proper curvature is created using a truing procedure. Subsequently, in the process of dressing, a portion of the bond is eliminated to reveal the sharp cutting edges of the abrasive grains. These two steps are executed in accordance with the instructions provided by the manufacturer. Table 2 displays the truing and dressing parameters.

Table 2. Truing and dressing parameters.

Parameter	Truing (Green Corundum)		Dressing (White Aluminium Oxide)	
Relative wheel speed	100 m/min		50 m/min	
Cutting direction	Down Cut		Down Cut	
Cutting depth (a_e)	10 µm		5 µm	
Feed rate (v_w)	20 mm/min		10 mm/min	
Grinding wheel	#1200 (D9)	#3000 (D3)	#1200 (D9)	#3000 (D3)
Truing/Dressing wheel	#600	#600	#320	#600

The grinding trials are conducted using a flat surface grinding technique. In this scenario, the grinding motion is performed in a raster pattern. In this experiment, it is important to highlight that the relative motion between the workpiece and the grinding tool is produced by the movement of the workpiece itself, with a velocity denoted as v_w, along the x and y machine axes. The movement of the grinding tool (z machine axis) only determines the cutting depth, represented by a_e. The raster movement in these trials begins in the upper right corner of the machining surface. The cutting tool moves towards the surface of the sample in a linear link motion. A second lead-in motion decreases the velocity of the link motion and guarantees a seamless and accurate transition to the grinding motion and additionally minimizes the tool load in transit. The grinding movement, which involves the workpiece moving in the positive x direction, follows a straight-line trajectory. After this, there is a further movement known as a lead-out and link. The grinding tool is moved to the right-hand side of the sample using a linear displacement motion. The workpiece undergoes motion in the negative x-direction and positive y-direction. Using this kinematic configuration, the vertical displacement of the glassy carbon sample results in the grinding path distance, indicated as a_p. This process is iterated until a segment of the sample is fully ground. The combination of the grinding wheel's direction and rotating cutting velocity v_c along with the movement of the workpiece, leads to cross-axis kinematics with a down-cutting direction. The generation of the NC code is facilitated by the utilization of NanoCam4 (Version 4.2309.9) and is visualized in Figure 3.

Based on the application as mould material for precision glass moulding, the grinding parameters were chosen to investigate the influence of tool properties and process control variables. The parameters are based on a cross-axis rotational process which is commonly used in mould manufacturing for optics. Further, the chosen parameters are additionally based on investigations from the literature [40–42]. Table 3 gives an overview of the investigated parameters, which includes tools with different diamond grain sizes D3 and D9 and varying cutting speed and feed rates. Other parameters are kept constant, like grain concentration, cutting depth, and path distance.

Table 3. Grinding process parameters.

	Parameter	Unit	Value
Kinematic	Cross-Axis	-	-
Tool properties	Diamond grain size/ Mesh size	[µm]	D3/#3000, D9/#1200
	Concentration	-	C140
	Bonding type	Hybrid (metal-resin)	-
Process parameters	Cutting speed	[m/s]	10, 20, 30
	Feed rate	[mm/min]	0.5, 100
	Cutting depth	[µm]	1
	Path distance	[mm]	0.01

3. Results and Discussion

3.1. Indentation and Scratch Investigation

To investigate the primary attributes of scratching on glassy carbon, a sequence of individual scratch experiments was performed and afterwards analysed utilizing scanning electron and atomic force microscopy. The behaviour of each scratch has a notable correlation with the extent of cutting depth. At reduced cutting depths, the scratch exhibits mostly ductile behaviour, as indicated by the substantial plastic deformation revealed in Figure 4a and beginning small minor additional scratches in the contact zone (Figure 4b). The surface of the scratch exhibits radial cracks that are inclined in a forward direction, resulting in a funnel-shaped formation (Figure 4c). The wider opening of this funnel formation is oriented towards the direction of the scratch mark's progression.

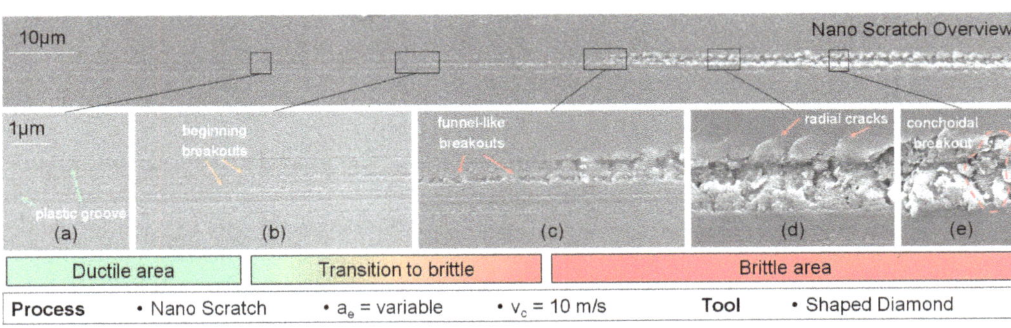

Figure 4. SEM pictures of nano-scratch behaviour from ductile to brittle (**a**–**e**).

As the magnitude of the depth of cut is increased, the occurrence of radial fractures propagates throughout the contact area at the outer margins of the ductile region (Figure 4c). An additional increase in the depth of the cut results in the generation of large beginning breakouts, accompanied by the presence of partially fragmented glassy carbon chips that stay inside the scratched area (Figure 4d). The radial fractures exhibit partial widening and form a beginning chip which is not removed. As a result, an increasing number of fragmented chips become apparent, and the expanded radial fractures give rise to bigger breakouts that extend to the farthest points of the radial cracks. After this stage, observable lateral fractures become evident, leading to the formation of clod-like breakouts that replace the funnel-shaped breakouts (Figure 4e). The distinguishing factor between funnel-shaped and clod-like breakouts lies in the character of their contact zones. The funnel-shaped breakouts are confined to the scratch region, but the clod-like breakouts exhibit larger dimensions compared to the chipping zone.

Moreover, by the analysis of the two-dimensional cross-section profiles of the scratch along the track, it is possible to analyse the plastic behaviour at different depths of cut as seen in Figure 5. In general, the resulting scratch depth is overproportioned to the real cutting depth of the tool, which indicates brittle behaviour at deeper scratch depths. In the context of purely plastic deformation at the beginning of the scratch, there are no disturbances or disruptions, and the absence of chip formation indicates that the plastic deformation is only characterized by compressive material alteration. In the scenario of emerging radial fractures, which starts on the second left AFM image in Figure 5, there is no observable material displacement occurring at the outer boundaries of the scratch track, which indicates that the crack propagation is probably further outside than the plastic deformation area. The elevation of quantifiable material along the scratch is observed exclusively after the formation of radial fractures at the beginning scratch width of 1.27 μm and scratch depth of 9–10 nm, which is already in the range of brittle behaviour.

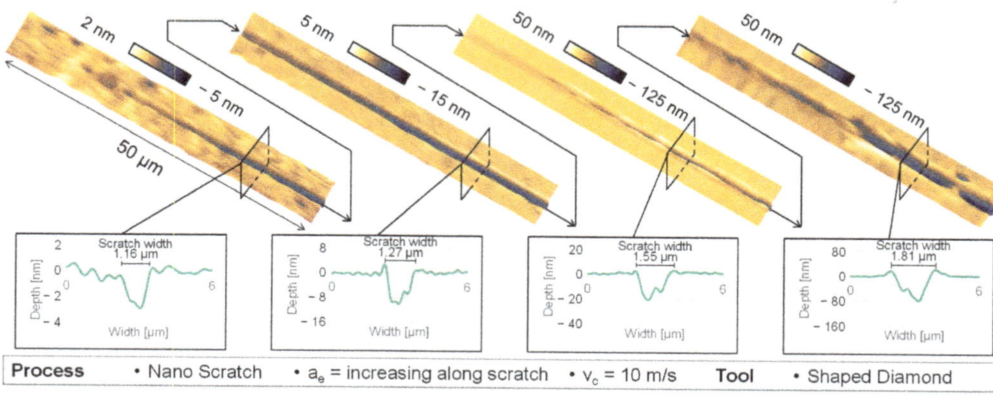

Figure 5. AFM and cross-section measurement of nano-scratch behaviour.

An additional analysis of the scratch marks produced at varying cutting speeds reveals that the material's behaviour remains consistent (Figure 6). However, it is seen that the formation of radial cracks is more inclined in the cutting direction, leading to clod-like breakouts that cover a narrower angle. This indicates that the crack propagation and material removal can be controlled with the cutting speed and it does show deterministic behaviour similar to the general ductile and brittle behaviour with increasing cutting depth.

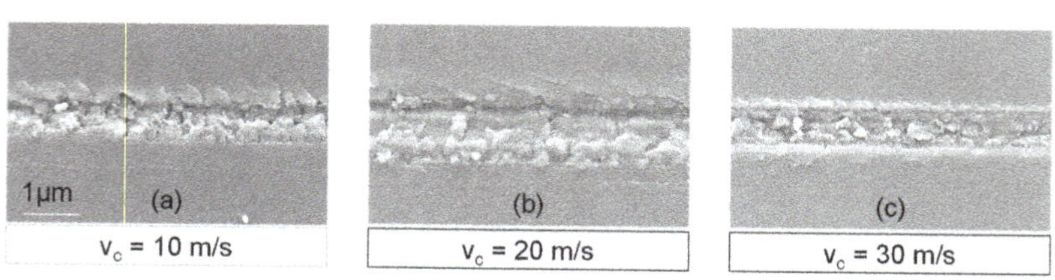

Figure 6. Visualization of radial crack inclination in varying cutting speeds. v_c = 10 m/s (**a**). v_c = 20 m/s (**b**). v_c = 30 m/s (**c**)

Further examination of the depth of scratches along the path of the scratch verifies the previously discussed phenomenon, as seen in Figure 7. The observed scratch depths were adjusted in a manner that aligns with the initiation of the brittle zone when there is a notable rise in scratch depth. It is worth noting that during the transition to the brittle-hard zone, there is a substantial rise in fracture depth, which can be explained by the breakouts and crack propagation exceeding the cutting depth of the diamond. At a cutting speed of 30 m/s, there is evidence of a notable decrease in fracture depth inside the brittle region compared to lower cutting speeds. This indicates a reduction in the propagation of cracks at greater depths, which in turn facilitates the early emergence of lateral cracks. Furthermore, there is a noticeable variation in depth within the ductile area. When the cutting speed is set to 10 m/s, it is noticed that the plastic zone exhibits increased length and depth. At lower cutting speeds, the plastic deformation strain is increased, leading to the occurrence of incipient separation fractures at a later stage. The region exhibiting exclusively ductile behaviour becomes apparent when the cutting speed is set at 10 m/s, resulting in a scratch depth of 5 nm. Similarly, with cutting velocities of 20 and 30 m/s, the highest ductile scratch depth seen is limited to 3 nm.

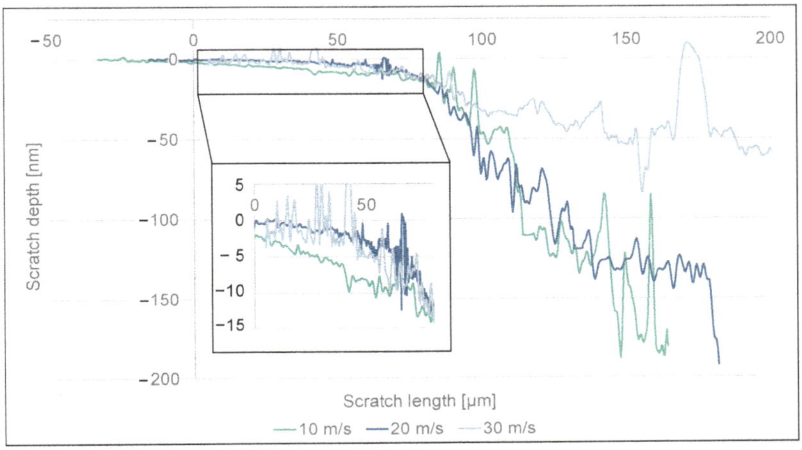

Figure 7. Scratch depth along the scratch length with varying cutting speeds.

To further examine the fracture behaviour in the subsurface zone, an additional scratch track was created with equidistant cutting depth, resulting in brittle breakouts. Subsequently, the subsurface area was removed using sphere calotte grinding to analyse the radial and lateral crack behaviour in different depths below the surface as shown in Figure 8. The presence of radial fractures is predominantly seen to be arranged either perpendicular or at an oblique angle relative to the cutting direction. In certain instances, the radial fractures exhibit a flared angle. The clod and shell-like breakouts exhibit shallower depths compared to the radial fractures and are typically situated posterior to a radial crack along the direction of the section. The conchoidal fractures, also known as lateral fractures, occur when a crack originates at a minor angle to the surface and extends until the chip breaks away entirely, causing the crack to propagate back towards the surface.

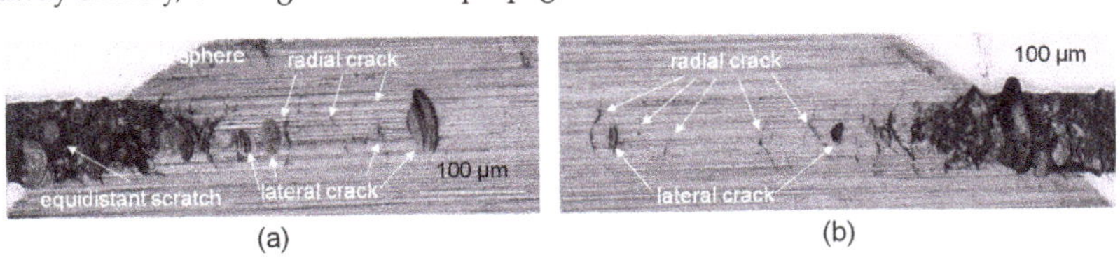

Figure 8. Subsurface investigation of nano-scratch revealed by calotte grinding. Subsurface damage at beginning of scratch (**a**) and end of scratch (**b**).

Additional indenter examinations including varying magnitudes of the Vickers pyramid forces revealed an absence of observable fractures, with only an evident indication of plastic behaviour (Figure 9a). At lower levels of applied indentation forces, the absence of any apparent indentation can be identified as the elastic response shown by the material. In addition, it can be observed that there were no radial cracks observed around the edge of the pyramid, indicating a characteristic of low residual stress. Similar to the subsurface crack propagation of a single scratch, the same pattern of activity is also observable in the indentation experiments at higher indentation forces. The clod-like breakouts that may be observed are apparent along the sides of the Vickers pyramid when viewed from the top (Figure 9b). Upon seeing the subsurface through a cross-sectional view, it becomes apparent that the edge zone has radial fractures that diverge from the sides of the pyramid and extend laterally deeper inside the material (Figure 9c,d). The observed phenomenon aligns with the conchoidal breakouts identified during the subsurface scratch test, as they progressively extend horizontally into the subsurface zone until reaching the removal of the chip. Likewise, the presence of plastic deformation in the central region is apparent, as shown by the observable Vickers pyramid. Subsequent examinations including varying magnitudes of the Vickers pyramid forces revealed an absence of observable fractures, with only an evident indication of plastic behaviour (Figure 9a). At lower levels of applied indentation forces, the absence of any apparent indentation can be identified as the elastic response shown by the material. In addition, it can be observed that there were no radial cracks observed around the edge of the pyramid, indicating a characteristic of low residual stress.

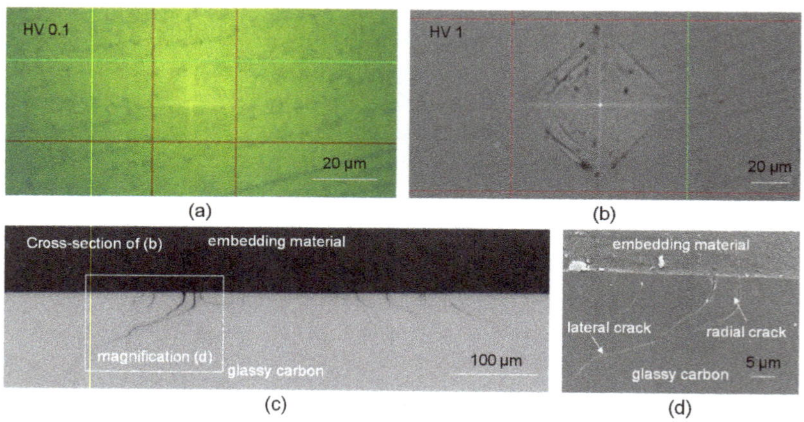

Figure 9. Vickers indentation crack behaviour and cross-section view. HV0.1 (**a**). HV1 (**b**). Cross-section view of HV1 indentation (**c**,**d**).

3.2. Grinding Investigation

A study was conducted to investigate the grinding of glassy carbon, as scratch tests do not consider the interaction between several diamond grains in contact. The inquiry focused on analysing the material's behaviour through the manipulation of process parameters, including the cutting speed, feed rate, and diamond grain size of the tool. In order to establish a precise and objective comparison, an individual variable was deliberately altered and then evaluated for comparison. Statistical reliability was established by conducting a minimum of three tests for each parameter set. The test specimens were subsequently analysed for surface topography using a confocal microscope, scanning electron microscope, and atomic force microscope. An analysis of surface roughness was conducted to determine a relationship between process parameters and surface structure. Topographical imagery can aid in the verification of meteorological results. An analysis of roughness metrics Sa and Sq at cutting speeds of 10, 20, and 30 m/s demonstrates a noticeable pattern of heightened roughness with higher cutting speeds. The visual representation of this discovery can be seen in Figure 10. Further grinding wheels with an average diamond grain diameter of 3 µm (D3) have lower surface roughness compared to grinding wheels with a relatively larger average diamond grain size of 9 µm (D9). This supports the common belief that an increase in grain size leads to thicker chips, resulting in larger breakouts and more brittle behaviour. Figure 10 demonstrates the impact of increasing roughness by a factor of more than two, from D3 to D9, on grain size at a velocity of 30 m/s.

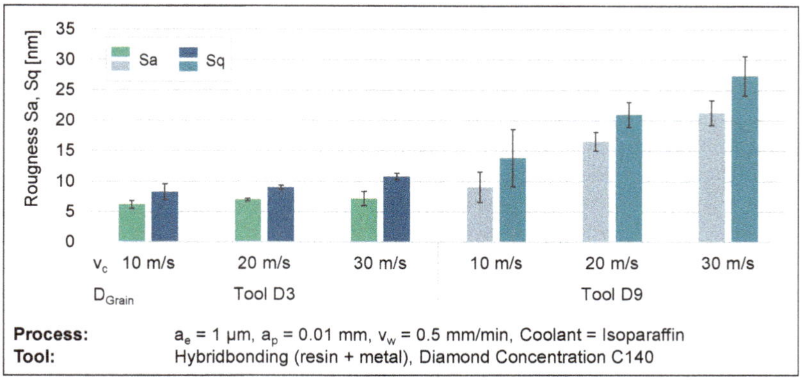

Figure 10. Surface roughness results from varying cutting speeds and diamond grain sizes.

This discovery contradicts the widely known assumption in the literature that greater cutting speeds lead to smaller chip removal and therefore result in improved roughness values [43–47]. Upon closer analysis of the surface topography depicted in Figure 11, it becomes apparent that increasing the cutting speed does indeed lead to an increase in surface roughness. Nevertheless, as the cutting speed is increased, the surface shows disruptions and a higher frequency of imperfections or slight chipping. At higher cutting speeds, the individual paths become more noticeable due to an increased accumulation of material. Upon closer examination, it is observed that the pile-up structures at higher cutting speeds exceed the anticipated kinematic roughness. This phenomenon cannot be accounted for by the plastic deformation of the material. Hence, the observable accumulation formations are formed due to the tool wear of the cutting tool, which accounts for the rise in surface unevenness at elevated cutting velocities.

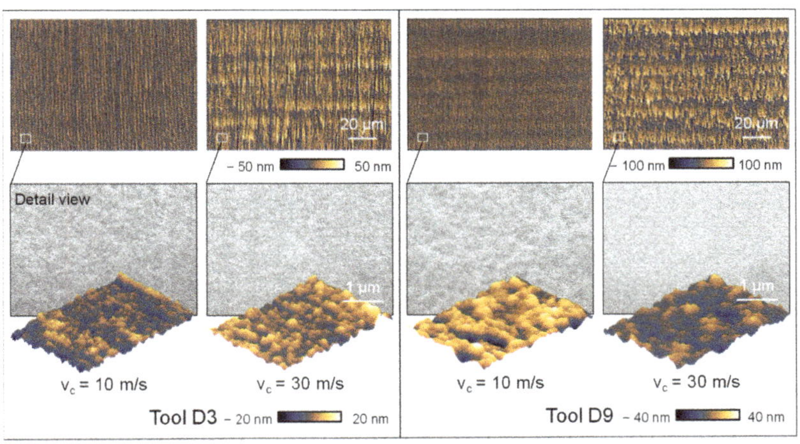

Figure 11. Topographic images and detail view of varying cutting speeds and diamond grain sizes.

Upon analysing the grinding topography observed in the SEM and AFM images, higher cutting speeds result in the formation of thinner and more distinct grinding marks caused by the rotation of the grinding wheel. This observation is further confirmed by the thorough analysis of the surface topographies. Increasing the cutting speed creates a surface topography characterized by less chipping and a more homogeneous visual aspect. The validity of this assertion is confirmed by the utilization of atomic force microscopy (AFM) to measure the roughness of the surface. This is demonstrated in Figure 12, which showcases the findings acquired through the utilization of the D3 grinding wheel as a specific example. The AFM images show a mostly consistent surface; however, at a velocity of $v_c = 10$ m/s, the surface displays the presence of overlapping irregular globular breakouts. Enhancing the cutting speed leads to a significant diminishment in the dimensions of globular breakouts, together with a decline in their depth, as confirmed by the AFM roughness measurements. Hence, it can be determined that a higher cutting speed leads to an improvement in the surface roughness. The presence of tool wear, caused by the elevated cutting speeds, shows a more pronounced influence on the resultant roughness than the cutting speed alone.

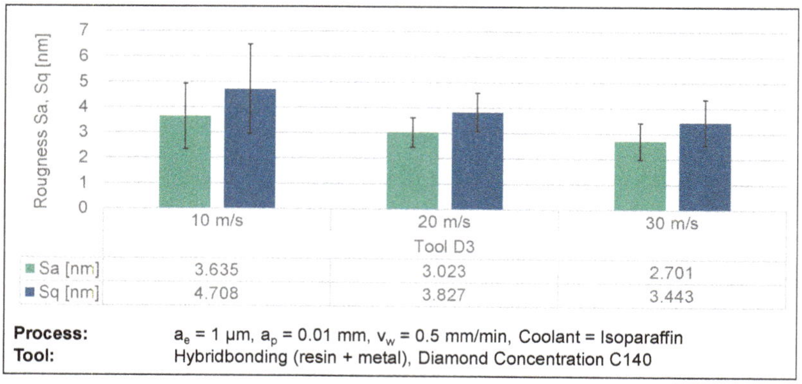

Figure 12. AFM surface roughness results from varying cutting speeds.

The results of scratch tests indicate that increasing the cutting speed causes the funnel-like and conchoidal breakouts to occur in a narrower separation fracture. This finding further supports the idea that higher cutting speeds result in lesser surface roughness. Nevertheless, there are no discernible directional structures, such as funnel shapes or other patterns, that may be observed on the ground surface. The absence of uniform contact conditions among the abrasive grains leads to unregulated breakouts, indicating the occurrence of fracture propagation in both perpendicular and parallel cutting directions. However, breakouts at higher cutting speeds exhibit a reduced form factor, which improves the surface roughness. The fractures, which are less than 1 µm in size, indicate that the material removal occurs in the contact zone between the diamond grain (3 µm and 9 µm) and the glassy carbon. This implies a material removal behaviour within the transition zone from ductile to brittle, with no apparent radial surface cracks present (Figure 4b,c).

Further examination of the impact of the feed rate reveals contrasting patterns in relation to the diamond grit sizes used while maintaining a constant cutting speed of $v_c = 30$ m/s. An improvement in roughness is observed with an increase in feed rate for smaller diamond grit sizes (D3), whereas a rise in surface roughness is noted for larger diamond grit sizes (D9). This phenomenon is seen in the roughness values depicted in Figure 13.

The examination of the topography images validates the observed roughness values, as it is seen that both the macroscopic roughness and the microstructure exhibit consistency. The accepted hypothesis regarding the D9 grinding wheel suggests that there is a positive correlation between the feed rate and the chip thickness, thus resulting in higher roughness when increasing the feed rate. The D3 grinding wheel exhibits an inverse behaviour. The microstructure (D3) seen at a feed rate of $v_W = 100$ mm/min has a high degree of homogeneity on its surface, characterized by the presence of isolated break-out formations of minimal size, as depicted in Figure 14. Conversely, an observable phenomenon of elongated fractures in the grinding direction becomes apparent when the feed rate increases on the D9 grinding wheel. Furthermore, the surface on both grain sizes exhibits an oscillating structure, likely attributed to the machine tool's control mechanism. Nevertheless, it has no substantial effect on the roughness measurements.

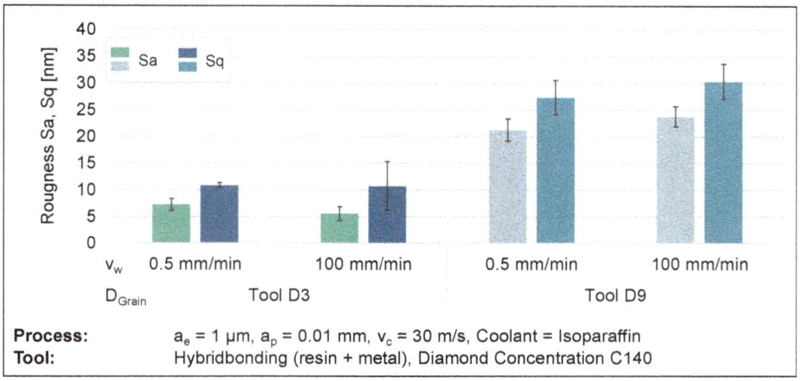

Figure 13. Surface roughness results from varying feed rates and diamond grain sizes.

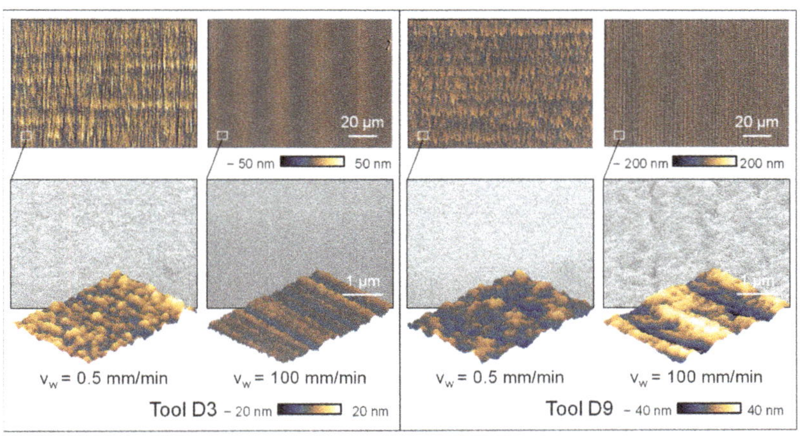

Figure 14. Topographic images and detail views of feed rates and diamond grain sizes.

The improvement of surface topography achieved through the utilization of a D3 grinding wheel may be defined by the impact of the bonding agent on the grinding tool. The grinding wheel bond exhibits a more uniform distribution when the concentration of diamonds remains constant, but the average diamond grain diameter decreases. This phenomenon facilitates increased deformation of the grinding wheel, resulting in elevated surface pressure on the material being ground. This phenomenon promotes the plastic deformation of glassy carbon and prevents or limits the spread of cracks. In the context of the D9 grinding wheel, apart from the unique behaviour resulting from the bond, the increased chip thickness can also induce greater brittleness. Higher feed rates evidently result in more visible grinding marks on the direction of the rotating grinding wheel. The grinding marks also indicate more ductile grinding behaviour in the case of a D3 tool as seen in the nano-scratch investigations (Figure 4b). Additionally, the AFM image of the close-up shows a similar pattern at v_W = 100 mm/min as investigated in the nano-scratch test and indicates ductile behaviour with plastic flow, because of the visible grinding marks. In the case of D9 tools, the overall removal mechanism shows brittle behaviour, most likely due to the more dominant effect of increased chip thickness and wider grinding marks based on the higher diamond grain size.

Additionally, the occurrence of pile-ups in the feed rate direction are not evident which indicates less tool wear due to the overall short use of the grinding wheels with feed rates

v_W = 100 mm/min and which supports the aforementioned influence of the tool wear at lower feed rates.

The findings indicate that the material response of glassy carbon during ultra-precision grinding is characterized by a high degree of complexity, with many factors typically interacting simultaneously. The scratch tests facilitated the identification of fundamental material behaviour, hence providing some insight into the grinding process. To enhance comprehension, Figure 15 was utilized to reconstruct the crack propagation seen during scratch testing, based on the obtained data.

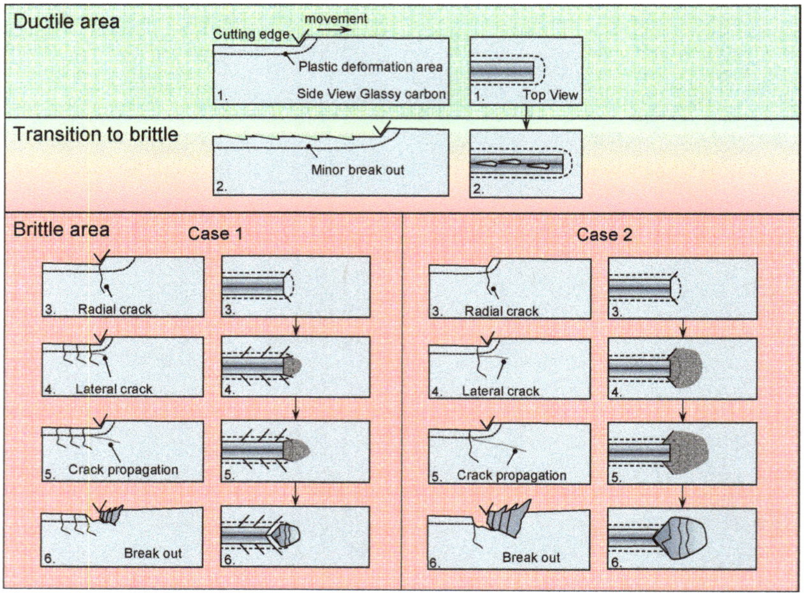

Figure 15. Crack propagation and breakout behaviour of glassy carbon. Increasing cutting depth from 1–6. Only plastic deformation (ductile) 1. Beginning break outs (transition) 2. Crack behaviour (brittle) 3–6.

Once the elastic behaviour of glassy carbon is surpassed, the material undergoes plastic deformation and compression. Once the plastic deformation capacity is surpassed, the initial fractures to emerge are minor funnel-like breakouts that probably have their initiation due to a radial crack which transforms into a lateral crack. The brittle behaviour shows two cases which differ in the area of effect. Both cases start with radial cracks that are orthogonal to the cutting direction and exceed the plastic deformations area. As the cutting edge progresses, a lateral fracture emerges, originating at an elevated position due to the surpassing of shear forces over the tensile strain in this region. The convergence of the lateral crack and radial fractures culminates in the formation of a conchoidal breakout. In case 1, the breakout is not formed by the outmost radial cracks compared to case 2. In the general case of brittle-hard behaviour, the radial cracks exhibit crack propagation that extends beyond the depth of cut. In the context of the ductile zone, it is unlikely that fracture propagation extends beyond the plastic region. The confirmation of this phenomenon may be obtained by the examination of the grinding process, where mostly globular chipping is observed. This chipping does not exhibit any discernible fracture development that follows a certain path. Furthermore, the presence of a consistently identifiable structure provides evidence for a machining process that operates within the ductile range.

4. Conclusions

In conclusion, this study could demonstrate the general behaviour of glassy carbon under abrasive machining conditions. Specifically, the fracture behaviour may be systematically studied and comprehensively explained. The findings obtained from the scratch experiments may be used for the grinding process, therefore providing a basis for understanding potential material behaviour. The nano-scratching process initially exhibits ductile plastic deformation, which is then followed by the occurrence of funnel-shaped breakouts in the contact zone as the cutting depth increases. Additionally, increased cutting depth results in mostly brittle material behaviour characterized by the initiation of radial fractures that then propagate into lateral cracks, resulting in the formation of conchoidal breakouts. The impact of process parameters and other tool-related parameters induced by grinding has distinct effects on the final surface roughness and topography. Through the extraction of microtopography, it can be demonstrated that the overall predicted grinding behaviour aligns with the general assumptions presented in the literature. Increasing the cutting speed while using cross-axis kinematic grinding leads to improved surface roughness. Conversely, increasing the diamond grain sizes and feed rates have the opposite effect, causing an increase in surface roughness. The primary factor influencing the roughness is the size of the diamond grains, whereas the impact of cutting speed and feed rate is minor. In addition, roughness values of less than 5 nm can be achieved with homogeneous surface topography, which makes it possible to use glassy carbon as a mould material. Nevertheless, the tool wear has a substantial influence on the roughness of the surface when using hybrid bonding (metal-resin) tools at reduced feed rates. The decrease in tool diameter can result in pile-up structures that exceed the theoretical kinematic roughness. This phenomenon bears great significance in rotational grinding processes, which are frequently employed in the fabrication of optical lenses. Typically, these processes include operating at low rotating speeds while cutting the central region of the optical mould surface. Hence, the increased surface roughness might restrict the use of glassy carbon moulds. The potential for ultra-precision ground glassy carbon moulds is significant when using other three-axis ground surfaces with constant greater feed rates, such as moulds for inclined cover glasses.

In addition to tool wear, the bonding agent itself has a beneficial effect on the surface roughness when higher feed rates are used in combination with small diamond grit sizes. This supports the theory that the resulting compressive stress leads to more ductile behaviour, as it reduces the ability of cracks to propagate and initiate.

However, the researchers also suggest that it is crucial to undertake further investigations, such as subsurface zone analyses and modifications of additional process parameters, in order to fully comprehend the material behaviour in ultra-precision grinding. Conducting research on process parameters involving increased cutting speed and feed rates, while simultaneously monitoring the effects of tool wear, might be beneficial in utilizing glassy carbon as a material for moulds.

These additional studies can provide new insights into the behaviour of separation fractures and the process of ductile machining. Furthermore, the impact on precision glass moulding with glassy carbon based on the resulting surface roughness induced by grinding has not yet been investigated.

Author Contributions: Conceptualization, K.J.; methodology, K.J. and D.P.W.; validation, K.J.; investigation, K.J., R.M. and D.P.W.; writing—original draft preparation, K.J.; writing—review and editing, T.G. and T.B.; visualization, K.J. and D.P.W. All authors have read and agreed to the published version of the manuscript.

Funding: This research received no external funding.

Data Availability Statement: Data is contained within the article.

Conflicts of Interest: The authors declare no conflict of interest.

References

1. Stenchly, V.; Reinert, W.; Quenzer, H.-J. Modular packaging concept for MEMS and MOEMS. *J. Phys. Conf. Ser.* **2017**, *922*, 12015. [CrossRef]
2. Yi, A.Y.; Jain, A. Compression Molding of Aspherical Glass Lenses-A Combined Experimental and Numerical Analysis. *J. Am. Ceram. Soc.* **2005**, *88*, 579–586. [CrossRef]
3. Klocke, F.; Dambon, O.; Yi, A.Y.; Wang, F.; Hünten, M.; Georgiadis, K.; Hollstegge, D.; Dukwen, J. Process Chain for the Replication of Complex Optical Glass Components. In *Fabrication of Complex Optical Components: From Mold Design to Product*; Brinksmeier, E., Gläbe, R., Riemer, O., Eds.; Springer: Berlin/Heidelberg, Germany, 2013; pp. 119–132. ISBN 978-3-642-33000-1.
4. Zhang, L.; Liu, W. Precision glass molding: Toward an optimal fabrication of optical lenses. *Front. Mech. Eng.* **2017**, *12*, 3–17. [CrossRef]
5. Dukwen, J.; Friedrichs, M.; Liu, G.; Tang, M.; Dambon, O.; Klocke, F. Tribological wear analysis and numerical lifetime prediction of glassy carbon tools in fused silica molding. *Wear* **2016**, *364–365*, 144–153. [CrossRef]
6. Grunwald, T.; Wilhelm, D.P.; Dambon, O.; Bergs, T. Influence of Glassy Carbon Surface Finishing on Its Wear Behavior during Precision Glass Moulding of Fused Silica. *Materials* **2019**, *12*, 692. [CrossRef]
7. Brinksmeier, E.; Mutlugünes, Y.; Klocke, F.; Aurich, J.C.; Shore, P.; Ohmori, H. Ultra-precision grinding. *CIRP Ann.* **2010**, *59*, 652–671. [CrossRef]
8. Bifano, T.G.; Dow, T.A.; Scattergood, R.O. Ductile-Regime Grinding: A New Technology for Machining Brittle Materials. *J. Eng. Ind.* **1991**, *113*, 184–189. [CrossRef]
9. Huang, H.; Li, X.; Mu, D.; Lawn, B.R. Science and art of ductile grinding of brittle solids. *Int. J. Mach. Tools Manuf.* **2021**, *161*, 103675. [CrossRef]
10. Hübner, K.-H. Anorganische nichtmetallische Werkstoffe. In *Werkstoffkunde*; Springer: Berlin/Heidelberg, Germany, 2018; pp. 357–395.
11. Dübgen, R.; Popp, G. Glasartiger Kohlenstoff Sigradur®—Ein Werkstoff für Chemie und Technik. *Mat. Wiss. Werkstofftech.* **1984**, *15*, 331–338. [CrossRef]
12. Harris, P.J.F. Fullerene-related structure of commercial glassy carbons. *Philos. Mag.* **2004**, *84*, 3159–3167. [CrossRef]
13. Jurkiewicz, K.; Duber, S.; Fischer, H.E.; Burian, A. Modelling of glass-like carbon structure and its experimental verification by neutron and X-ray diffraction. *J. Appl. Crystallogr.* **2017**, *50*, 36–48. [CrossRef]
14. Sharma, S. Glassy Carbon: A Promising Material for Micro- and Nanomanufacturing. *Materials* **2018**, *11*, 1857. [CrossRef]
15. Grunwald, T. *Modellierung des Werkzeugverschleißes bei der Quarzglasumformung*, 1st ed.; Apprimus Wissenschaftsverlag: Aachen, Germany, 2021; ISBN 9783863599812.
16. HTW Germany. Available online: http://www.htw-germany.com/technology.php5?lang=en&nav0=2&nav1=16 (accessed on 20 October 2023).
17. SCHOTT N-BK7® | SCHOTT Advanced Optics. Available online: https://www.schott.com/shop/advanced-optics/de/Optisches-Glas/SCHOTT-N-BK7/c/glass-SCHOTT%20N-BK7%C2%AE (accessed on 20 October 2023).
18. Bansal, N.P.; Doremus, R.H. *Handbook of Glass Properties*; Academic Press Inc.: Orlando, FL, USA, 1986; ISBN 9780080523767.
19. *E8/E8M-22*; Test Methods for Tension Testing of Metallic Materials. ASTM International: West Conshohocken, PA, USA, 2022.
20. Herbert, E.G.; Oliver, W.C.; Pharr, G.M. On the measurement of yield strength by spherical indentation. *Philos. Mag.* **2006**, *86*, 5521–5539. [CrossRef]
21. Goto, K.; Watanabe, I.; Ohmura, T. Determining suitable parameters for inverse estimation of plastic properties based on indentation marks. *Int. J. Plast.* **2019**, *116*, 81–90. [CrossRef]
22. Bucaille, J.L.; Stauss, S.; Felder, E.; Michler, J. Determination of plastic properties of metals by instrumented indentation using different sharp indenters. *Acta Mater.* **2003**, *51*, 1663–1678. [CrossRef]
23. Lawn, B.; Wilshaw, R. Indentation fracture: Principles and applications. *J. Mater. Sci.* **1975**, *10*, 1049–1081. [CrossRef]
24. Schinker, M.G.; Döll, W. *Abschlußbericht zum Forschungsvorhaben AIF Nr. 4613*; Fraunhofer-Inst. für Werkstoffmechanik: Freiburg, Germany, 1982.
25. Bellemare, S.C.; Dao, M.; Suresh, S. A new method for evaluating the plastic properties of materials through instrumented frictional sliding tests. *Acta Mater.* **2010**, *58*, 6385–6392. [CrossRef]
26. Sinha, S.K. 180 Years of scratch testing. *Tribol. Int.* **2006**, *39*, 61. [CrossRef]
27. Tabor, D. The physical meaning of indentation and scratch hardness. *Br. J. Appl. Phys.* **1956**, *7*, 159–166. [CrossRef]
28. Asqari, M.A.; Akbari, J. Investigation of cohesive FE modeling to predict crack depth during deep-scratching on optical glasses. *Ceram. Int.* **2018**, *44*, 16781–16790. [CrossRef]
29. Li, K.; Shapiro, Y.; Li, J. Scratch test of soda-lime glass. *Acta Mater.* **1998**, *46*, 5569–5578. [CrossRef]
30. Wang, W.; Yao, P.; Wang, J.; Huang, C.; Zhu, H.; Liu, H.; Zou, B.; Liu, Y. Controlled material removal mode and depth of micro cracks in precision grinding of fused silica—A theoretical model and experimental verification. *Ceram. Int.* **2017**, *43*, 11596–11609. [CrossRef]
31. Huang, H.; Lawn, B.R.; Cook, R.F.; Marshall, D.B. Critique of materials-based models of ductile machining in brittle solids. *J. Am. Ceram. Soc.* **2020**, *103*, 6096–6100. [CrossRef]
32. Liu, W.; Deng, Z.; Shang, Y.; Wan, L. Effects of grinding parameters on surface quality in silicon nitride grinding. *Ceram. Int.* **2017**, *43*, 1571–1577. [CrossRef]

33. Li, P.; Jin, T.; Xiao, H.; Chen, Z.; Qu, M.; Dai, H.; Chen, S. Effects of wheel speed on surface/subsurface damage characteristics in grinding of glass-ceramics. *Ceram. Int.* **2020**, *46*, 17717–17728. [CrossRef]
34. Guo, S.; Lu, S.; Zhang, B.; Cheung, C.F. Surface integrity and material removal mechanisms in high-speed grinding of Al/SiCp metal matrix composites. *Int. J. Mach. Tools Manuf.* **2022**, *178*, 103906. [CrossRef]
35. Li, C.; Piao, Y.; Meng, B.; Hu, Y.; Li, L.; Zhang, F. Phase transition and plastic deformation mechanisms induced by self-rotating grinding of GaN single crystals. *Int. J. Mach. Tools Manuf.* **2022**, *172*, 103827. [CrossRef]
36. Li, X.; Fan, Z.; Huang, S.; Lu, M.; Huang, H. Brittle-to-ductile transition in nanoscratching of silicon and gallium arsenide using Berkovich and Conical tips. *Appl. Surf. Sci.* **2023**, *637*, 157934. [CrossRef]
37. Qu, S.; Yao, P.; Gong, Y.; Chu, D.; Yang, Y.; Li, C.; Wang, Z.; Zhang, X.; Hou, Y. Environmentally friendly grinding of C/SiCs using carbon nanofluid minimum quantity lubrication technology. *J. Clean. Prod.* **2022**, *366*, 132898. [CrossRef]
38. Li, C.; Piao, Y.; Zhang, F.; Zhang, Y.; Hu, Y.; Wang, Y. Understand anisotropy dependence of damage evolution and material removal during nanoscratch of MgF_2 single crystals. *Int. J. Extrem. Manuf.* **2023**, *5*, 15101. [CrossRef]
39. Flosky, C.; Riemer, O.; Brinksmeier, E. Ultra-precision grinding of ceramics with injection moulded tools. In Proceedings of the Euspen's 16th International Conference & Exhibition, Nottingham, UK, 30 May–3 June 2016.
40. Schönemann, L.; Adam, B.; Riemer, O.; Karpuschewski, B. Precision machining of glassy carbon. In Proceedings of the Euspen's 23rd International Conference & Exhibition, Copenhagen, Dennmark, 12–16 June 2023.
41. Beiring, P.; Yan, J. Ultrasonic vibration-assisted microgrinding of glassy carbon. *Proc. Inst. Mech. Eng. Part C J. Mech. Eng. Sci.* **2019**, *233*, 4165–4175. [CrossRef]
42. Hwang, Y.; Cha, D.-H.; Kim, J.-H.; Kim, H.-J. Ultraprecision Grinding of Glassy Carbon Core for Mold Press Lens. *J. Korean Soc. Precis. Eng.* **2012**, *29*, 261–265. [CrossRef]
43. Berger, D.; Althoff, M.; Rickens, K.; Heinzel, C.; Brinksmeier, E. Präzisionsschleifen mit groben Diamantkörnern*/Precision grinding with coarse diamond grains—Application characteristics of coarse-grained diamond grinding wheels. *Werkstattstech. Online* **2016**, *106*, 387–393. [CrossRef]
44. Li, S.; Wang, Z.; Wu, Y. Relationship between subsurface damage and surface roughness of optical materials in grinding and lapping processes. *J. Mater. Process. Technol.* **2008**, *205*, 34–41. [CrossRef]
45. Malkin, S. *Grinding Technology: Theory and Applications of Machining with Abrasives*, 2nd ed.; Industrial Press: New York, NY, USA, 2008; ISBN 9780831132477.
46. Sinhoff, V.R. Feinbearbeitung Optischer Gläser in der Kleinserie. Ph.D. Thesis, RWTH Aachen University, Aachen, Germany, 1997. Available online: https://books.google.de/books/about/Feinbearbeitung_optischer_Gl%C3%A4ser_in_der.html?id=4MIPXwAACAAJ&redir_esc=y (accessed on 10 December 2023).
47. Rickens, K. Deterministisches Schleifen Optischer Gläser mit Grobkörnigen Diamantschleifscheiben. Ph.D. Thesis, University Bremen, Bremen, Germany, 2010. Available online: https://www.shaker.de/de/content/catalogue/index.asp?lang=de&ID=8&ISBN=978-3-8322-9975-0 (accessed on 10 December 2023).

Disclaimer/Publisher's Note: The statements, opinions and data contained in all publications are solely those of the individual author(s) and contributor(s) and not of MDPI and/or the editor(s). MDPI and/or the editor(s) disclaim responsibility for any injury to people or property resulting from any ideas, methods, instructions or products referred to in the content.

micromachines

Article

Influence of Diamond Wire Saw Processing Parameters on the Sawn Surface Characteristics of Silicon Nitride Ceramics

Siyuan Zhang, Yufei Gao *, Xingchun Zhang and Yufeng Guo

Key Laboratory of High Efficiency and Clean Mechanical Manufacture of MOE, School of Mechanical Engineering, Shandong University, Jinan 250061, China
* Correspondence: yfgao@sdu.edu.cn

Abstract: For the slicing of superhard silicon nitride ceramics, diamond wire sawing technology has great potential for application, and its slicing surface characteristics are an important indicator of cutting quality. In this paper, the sawing experiments of silicon nitride ceramics were carried out within the range of industrial processing parameters of diamond wire sawing (saw wire speed: 800–1600 m/min, workpiece feed speed 0.1–0.4 mm/min). The effects of cutting parameters on the surface morphology, surface roughness and waviness of the as-sawn slices were analyzed. The results show that within the range of sawing parameters for industrial applications, the material on the diamond wire as-sawn surface of silicon nitride ceramics is removed mainly in a brittle mode, with the slice morphology showing brittle pits and regularly distributed wire marks in the 20–55 μm scale range. The surface roughness of the slices along the workpiece feed direction ranges from 0.27 to 0.38 μm and decreases with increasing saw wire speed and decreasing feed rate. The surface waviness ranges from 0.09 to 0.21 μm, which is in good agreement with the changing trend of the sliced-surface roughness. The results of the study provide an experimental reference for promoting the engineering application of diamond wire sawing technology to the processing of silicon nitride ceramic slices.

Keywords: silicon nitride ceramics; diamond wire saw slicing; surface morphology; surface roughness; waviness

1. Introduction

Silicon nitride ceramics (Si_3N_4) have high strength, high hardness, low coefficient of thermal expansion, high thermal conductivity, excellent thermal shock resistance and good oxidation resistance. Si_3N_4 is often used in the aerospace and military industry, mechanical engineering, communications, electronics, automotive, energy, chemical and biological fields. It is particularly promising for electronic packaging substrate applications, suitable for applications in complex and extreme environments [1–3].

The Si_3N_4 ceramics substrate preparation process mainly includes the following procedures: first, the Si_3N_4 powder and other mixtures are sintered with Si_3N_4 ceramics blocks by using several methods such as atmospheric pressure sintering, hot pressure sintering, reaction sintering, sintering reaction re-sintering, pneumatic pressure sintering, and hot isostatic pressure sintering. Then, after subsequent cutting to obtain ceramic slices, the commercial Si_3N_4 substrates are obtained through further grinding and polishing. Currently, the thickness of Si_3N_4 ceramics substrates for commercial use ranges from 0.3 to 0.6 mm. Slicing is the forming process of the substrate, which is related to the surface quality of the Si_3N_4 ceramics substrate and thus affects the subsequent processing costs. Si_3N_4 ceramics slices are mainly processed by wire electrical discharge machining (WEDM) and laser cutting [4–6]. WEDM wire cutting of hard-brittle materials has many applications, but when applied to non-conductive Si_3N_4 ceramics, requires the addition of an auxiliary electrode, resulting in a complex slicing process [7]. Laser cutting requires high-power lasers for cutting large-thickness workpieces, which limits the application of this technology for cutting large-size Si_3N_4 ceramics substrates. In addition, laser cutting causes relatively deep thermal damage layers on the surface and subsurface of the ceramic substrate and slice warpage. In recent years, diamond wire saw slicing technology has been widely used for slicing hard-brittle materials such as photovoltaic and semiconductor silicon crystals [8,9], silicon carbide crystals [10], sapphire [11,12], quartz [13] and magnetic materials [14] due to the advantages of its high sawing performance and small kerf loss. For the slicing processing of Si_3N_4 ceramics, the diamond wire saw slicing technology has great application potential.

Surface characterization is one of the most important concerns in machining such as cutting and grinding [15–17]. In diamond wire saw slicing, a large number of diamond abrasives on the surface of the diamond saw wire carry out a cutting action to remove material. Scholars have conducted a series of studies on the characteristics of the diamond wire saw processing, focusing mainly on the surface integrity of processing such as slice surface morphology, surface roughness and subsurface damage. The materials processed are mainly focused on silicon crystals. It was found that wire saw cutting parameters have a significant effect on as-sawn slice surface properties [16–21]. Yin et al. [16] and Liu et al. [17] found that high wire speeds and low workpiece feed speeds resulted in a better slice-surface quality in their experiments of cutting polysilicon using diamond saw wires with different diameters. Qiu et al. [18,19] found that increasing the saw wire speed and decreasing the feed rate are beneficial for improving the sawing performance and reducing the surface roughness of photovoltaic silicon wafers in multi-wire cutting. Costa et al. [20] found a similar pattern of effects when cutting polysilicon. Huang et al. [21] found that increasing the wire speed can increase the proportion of material ductile domain removal on the slice surface and reduced the slice surface waviness period. Liang et al. [22] analyzed the variation in cutting force during the sawing of monocrystalline silicon, which increased with the increase in workpiece feed and decreased with the increase in wire speed.

Diamond wire saw slicing of Si_3N_4 ceramics has great potential for development. Therefore, in this paper, the diamond wire sawing experiments on Si_3N_4 ceramics were carried out. The influence of the slicing process parameters (wire speed 800–1600 m/min and workpiece feed speed 0.1–0.4 mm/min) that are practically applied in the industry on the surface properties of as-sawn Si_3N_4 ceramics slices was investigated. The study was conducted mainly from the perspectives of slice-surface morphology, surface roughness and waviness. The results of this paper provide an experimental reference for promoting the application of diamond wire saw slicing technology to the processing of Si_3N_4 ceramics slices.

2. Experiment

2.1. Experimental Equipment and Methods

The experiment equipment uses a single wire reciprocating cutting machine, which is oriented to cut materials under industrial conditions. The main working part of the machine consists of spools, tensioning wheels, guide wheels, diamond saw wires, loading tray, and cutting fluid spout. The single wire reciprocating cutter maintains a certain tension by means of a tensioning wheel. The diamond saw wire is driven by the spool in reciprocating motion. The workpiece is fixed on the loading tray and driven to provide a feeding motion. Under the combined action of saw wire reciprocating motion and feeding motion, the cutting of Si_3N_4 ceramic block in this experiment is completed. Figures 1 and 2 are schematic diagrams of the actual appearance and processing principle of the equipment, respectively.

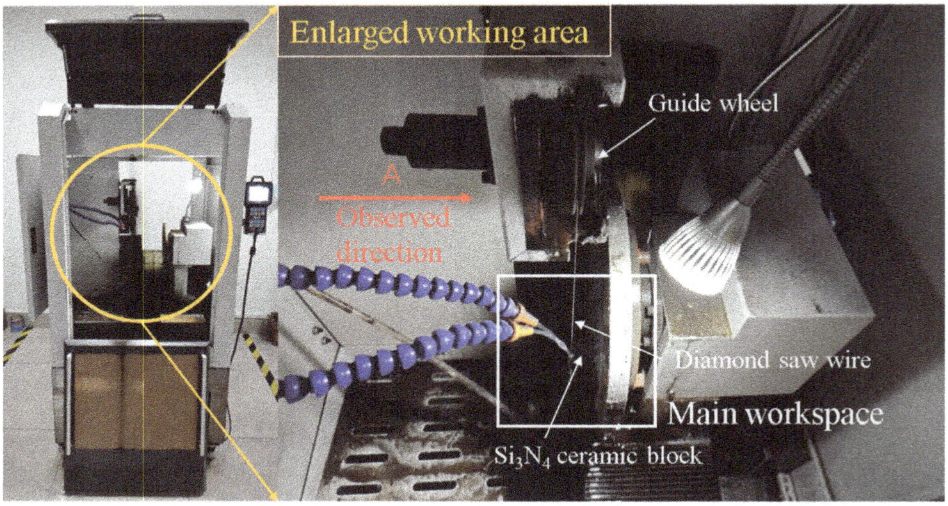

Figure 1. Experimental equipment apparatus.

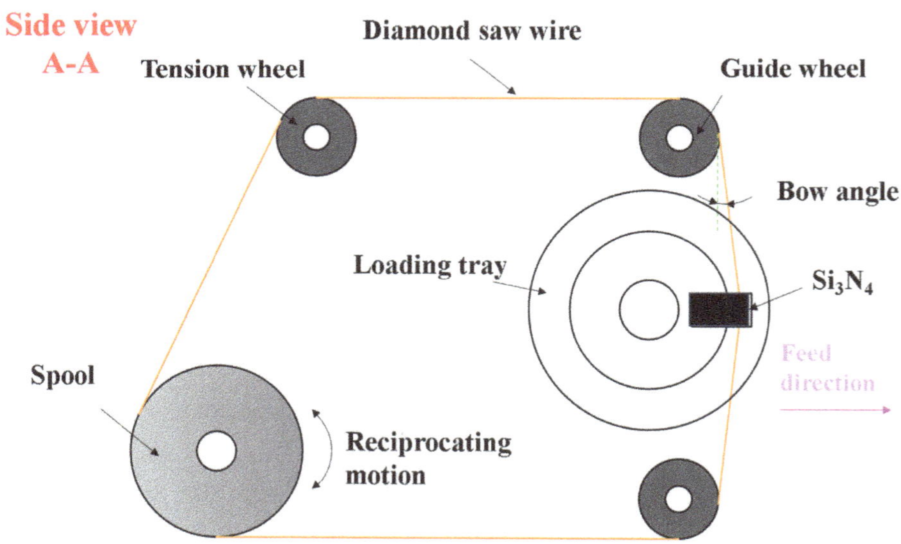

Figure 2. Schematic diagram of the wire saw cutting principle.

A 10 mm × 22 mm × 35 mm Si_3N_4 ceramic block was selected for this experiment. The workpiece was cut into slices of 1 mm thickness in the direction parallel to the 10 mm × 22 mm surface for detecting the sawn surface quality. The electroplated diamond saw wire was selected and its appearance is shown in Figure 3. The wear state of the processing tool affects the processing characteristics [23,24]. Therefore, the experiments were carried out in the steady state of diamond saw wire wear to avoid the influence of the tool wear on the surface properties of the slices. The parameters of the workpieces and saw wires used in the experiment are shown in Table 1. Pure water is used as cutting coolant.

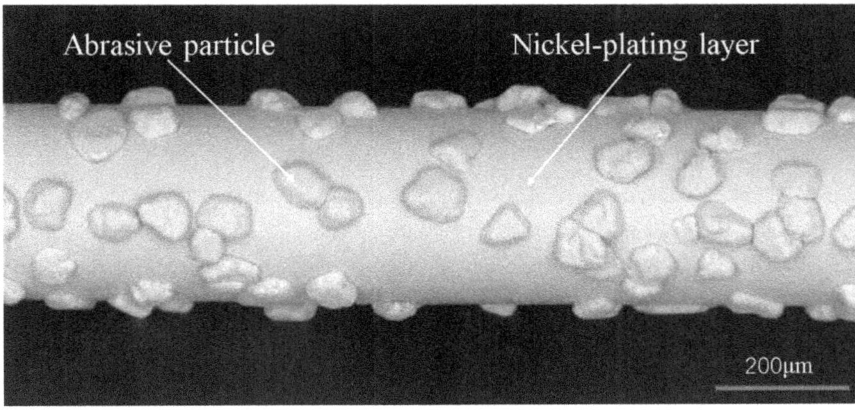

Figure 3. Diamond saw wire appearance.

Table 1. The parameters of saw wire and workpiece.

Parameter	Parameter Value
Workpiece size (mm)	10 × 22 × 35
Slice thickness (mm)	1
Diamond saw wire length (m)	80
Maximum saw wire envelope outer diameter (μm)	350
Type of abrasive grain	Surface nickel plated diamond (25% weight gain)
Abrasive grain size (μm)	30–40
Abrasive grain distribution density (grits/mm)	70–80

During the experiment, the diamond saw wire is wrapped onto the spool of the device and passes through the correct working position. Afterwards, it is repeatedly tensioned by means of a counterweight. Eventually the diamond wire is evenly wrapped onto the spool with the same degree of tension. The contact surfaces of the specimen and the carrier tray are fixed with two-component epoxy resin adhesive. The cutting experiment starts after checking the status of the cutting machine and the workpiece. The basic procedure from the cutting experiment to the slice quality inspection is as follows (as seen in Figure 4).

Wire speed and feed speed, as two of the important parameters of the wire sawing process, have a large impact on the machining process of hard-brittle materials. Therefore, a single-factor experiment was designed by varying the feed speed and wire speed separately during sawing. As shown in Table 2, a total of seven sets of cutting parameters were adopted for the two sets of experiments. Experiment 1 was set up as four control groups, and the effect of wire speed on the surface quality of as-sawn slices was observed by keeping the feed speed constant. In this case, a control group with a wire speed of 1600 m/min was used to study the machining quality under high speed conditions. Experiment 2 was also set up as four control groups. The wire speed was set as constant and the feed speed as variable. Qualitative analysis and quantitative measurements used the change in feed speed in relation to the change in surface quality.

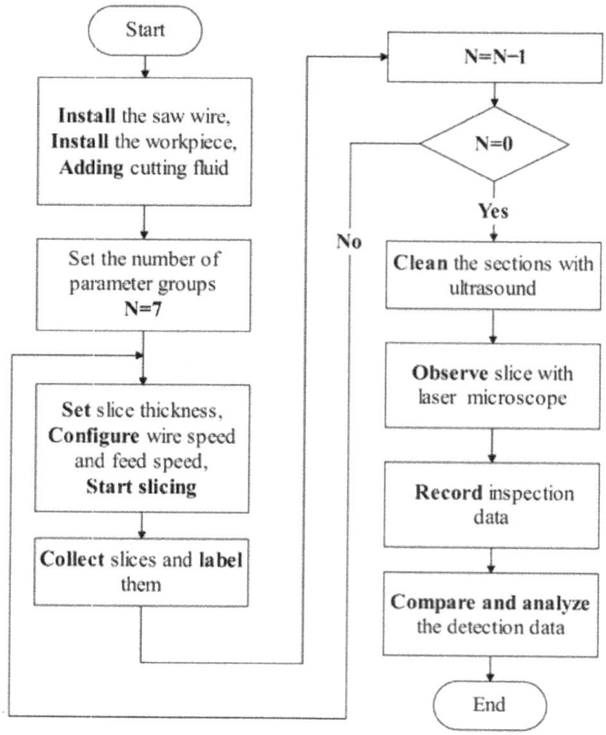

Figure 4. Flowchart of the experiment.

Table 2. Single-factor experiment table.

Processing Parameters	Wire Speed/m·min^{-1}	Feed Speed/mm·min^{-1}
Experiment 1	800 1000 1200 1600	0.3
Experiment 2	1200	0.1 0.2 0.3 0.4

2.2. *Evaluation of As-Sawn Surface Characteristics*

The evaluation of the as-sawn surface characteristics of Si_3N_4 ceramics slices will be approached from both qualitative and quantitative perspectives. The variation laws of as-sawn surface quality with process parameters and the data on the surface quality of the slices will be analyzed, respectively. Surface morphology characteristics, surface roughness, and waviness are important parameters that characterize machining quality [25,26], so these parameters are used to evaluate machining performance in experiments.

In order to exclude the effect of random errors on the experimental results, three as-sawn Si_3N_4 ceramics slices were cut and labeled under one set of cutting parameters. After ultrasonic cleaning of the slices, three sampling detection positions 1, 2 and 3 were selected as shown in Figure 5. According to the wire sawn morphological characteristics, the roughness value along the specimen feed direction is greater than that along the saw wire movement direction, and the waviness shows a certain pattern of variation along the specimen feed direction. So, the surface roughness and waviness of the Si_3N_4 ceramics slices were detected using a KEYENCE Laser Confocal Microscope made in Osaka, Japan along the specimen feed direction. The average value of the sampling data at the three sampling points is taken as the roughness or waviness value of the slice. The average of three slices sawn with the same parameter is taken as the final result. After all the data was collected, the analysis was performed to compare and study the changes in the surface quality of the slices at different wire speeds and feed speeds.

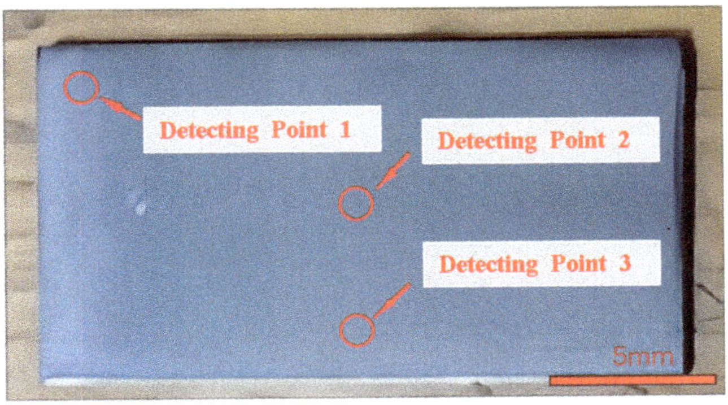

Figure 5. Schematic diagram of sampling detection location.

3. Results and Discussion

3.1. Surface Morphology Characteristics of As-Sawn Slices

Figures 6 and 7 show the surface morphology images of the slices taken using Laser Confocal Microscopy. As can be seen in Figure 6, under this set of process parameters, the formation of the sliced surface is mainly the result of brittle removal of the material, accompanied by a small number of ductile scratches. There are many brittle pits on the surface of the slices, which is due to the material removal form and characteristics of the cutting process of Si_3N_4 ceramics. On the one hand, Si_3N_4 ceramics are prone to brittle fracture when cutting, and the discontinuous chips result in brittle pits on the machined surface. On the other hand, due to the vibration of the saw wire in cutting [9,17] and the uneven size of the abrasive grains on the diamond saw wire surface [8], then the cut depths of abrasive grains are inconsistent at different positions of the wire surface, which results in an uneven surface. Numerous studies have shown that when the depth of abrasive cutting of hard-brittle materials is below the critical cutting depth, the material can be removed in a ductile manner to obtain high-quality machined surfaces [27,28]. The formation of larger-sized brittle pits left by material brittleness removal affects the serviceability of Si_3N_4 ceramics after cutting.

Combined with Figures 6 and 7, the surface of the Si_3N_4ceramics slices is also characterized by wire marks of varying depths, which are regularly distributed and periodically varied. Of these, Type 1 wire marks are slimmer in appearance but deeper in depth. The actual depth is basically between 2 and 4 μm, which is in the valley of the overall structural waveform. In contrast, Type 2 wire marks are thicker and more extensive but shallower in depth. The average depth of such wire marks is within 1 μm and is only a slight fluctuation in the overall structural waveform of the surface. Compared with the two types of wire marks, Type 1 wire marks have a greater influence on the surface quality of slices. Therefore, to improve the quality of Si_3N_4 ceramics in the wire saw cutting process, the first step is to reduce the depth of the Type 1 wire marks. That is, to reduce the range of height variation in the surface to reduce the degree of waviness and surface roughness.

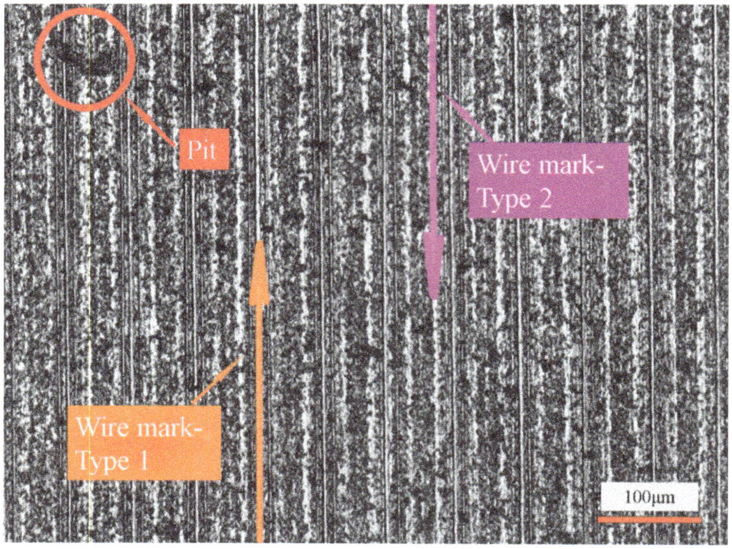

Figure 6. Basic morphology of sliced Si_3N_4 surface (wire speed 1000 m/min and feed speed 0.3 mm/min).

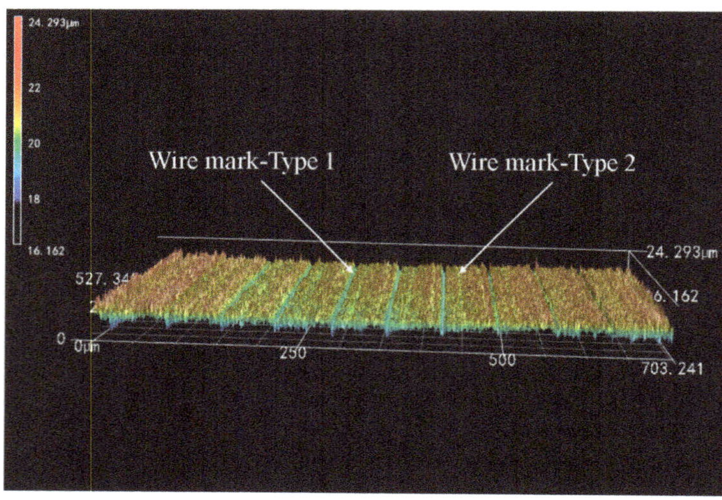

Figure 7. Observation of wire marks on the surface of Si_3N_4 slices (wire speed 1000 m/min and feed speed 0.3 mm/min).

Combined with the changes in the wire bow angle observed during machining, the cause of the periodic wire marks can be reasonably analyzed. Si_3N_4 is a highly hard-brittle material, which is difficult to machine in the production process. When the saw wire touches the workpiece, it is not actually able to quickly remove the excess material and realize the feed at the desired feed rate. As a result, the saw wire bends to create a bow angle. This increases the partial force in the feed direction. When the partial force in the feed direction is small, the wire saw cannot remove the workpiece material in time. It will cut back and forth within a small range of variation, resulting in Type 1 wire marks. When the bow angle increases to a critical point, the wire appears to "jump cut" the situation, along the feed direction to produce a faster jump. On the one hand, the width of the marks increases due to the faster feed. On the other hand, the depth of the marks decreases due to fewer reciprocating cuts in the same position. In this process, Type 2 wire marks appear. The alternation of these two processes during machining produces a periodically changing surface topographic feature.

3.1.1. Influence of Wire Speed on Slice Surface Morphology

Figure 8 shows the effect of the diamond wire speed on the slice surface morphology at a feed speed of 0.3 mm/min. With the increase in wire speed, the fluctuation of the surface shape of the workpiece decreases, and the wire marks become clearer gradually. The horizontal distance between the wire marks changes from about 66 μm to 51 μm. The wire marks spacing decreased by 22.7% over the course of the experiment. It is inferred that the density of wire marks is directly proportional to the wire speed in the same size inspection area. This also verifies the assertion made in the previous section. As the wire speed increases, the maximum bow angle produced at the same feed per unit time becomes smaller. The period of surface morphology change in the feed direction is also shortened. From the point of view of machining allowance removal, the change in brittle pits shows a certain trend. The density of brittle pits on the surface of the slices decreased during the increase in wire speed from 800 mm/min to 1600 mm/min. However, the brittle pits did not change significantly in the scale range.

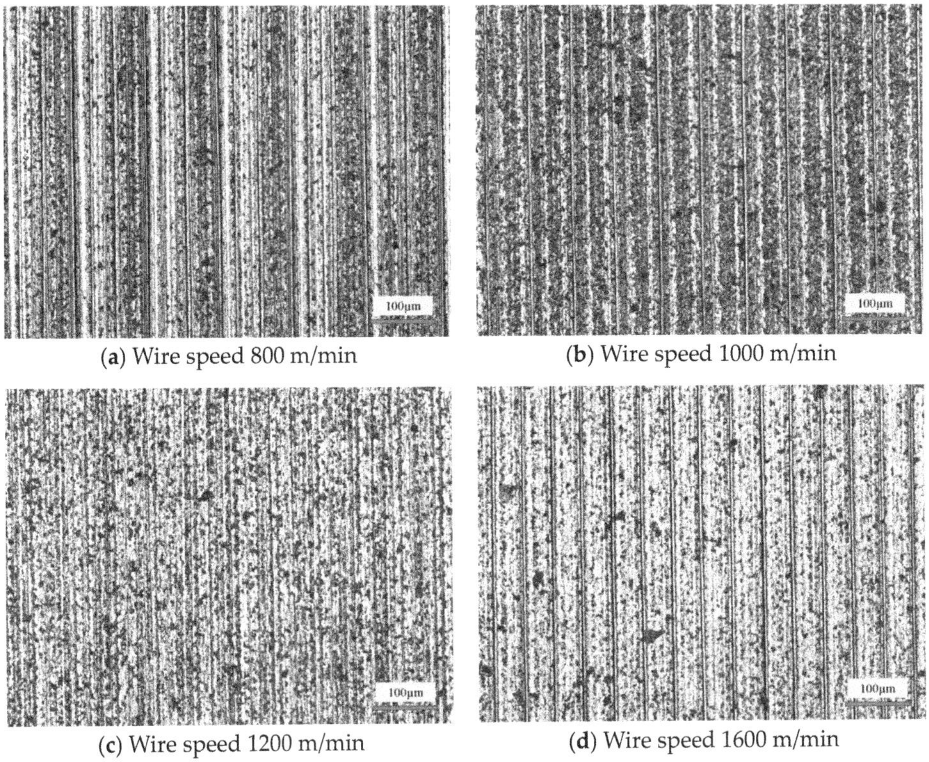

Figure 8. Surface morphology of slices at different wire speeds. (Feed speed 0.3 mm/min).

3.1.2. Influence of Feed Speed on Slice Surface Morphology

Figure 9 shows the effect of the workpiece feed speed on the slice surface morphology at a wire speed of 1200 m/min. Comparing the difference in surface morphology from

Figure 9a–d, it can be observed that the higher the feed speed, the higher the number of brittle pits. The area of a single brittle pit is getting larger and larger. The scale of the brittle pits on the sliced surface varies from 20 to 35 μm to the range of 30 to 55 μm. That is, the surface quality deteriorates as the feed speed increases. An increase in the workpiece feed speed increases the amount of feed per unit of time. The increased proportion of brittle removal in the whole material removal results in more brittle pits. In addition to this, the test process was observed under the condition of constant wire speed. It was found that the diamond saw wire bow angle became larger when the feed speed increased. This exerts a greater feed force on the workpiece, which increases the cut depth of the abrasive grains on the saw wire surface. The size of the pits during material brittle removal also becomes larger.

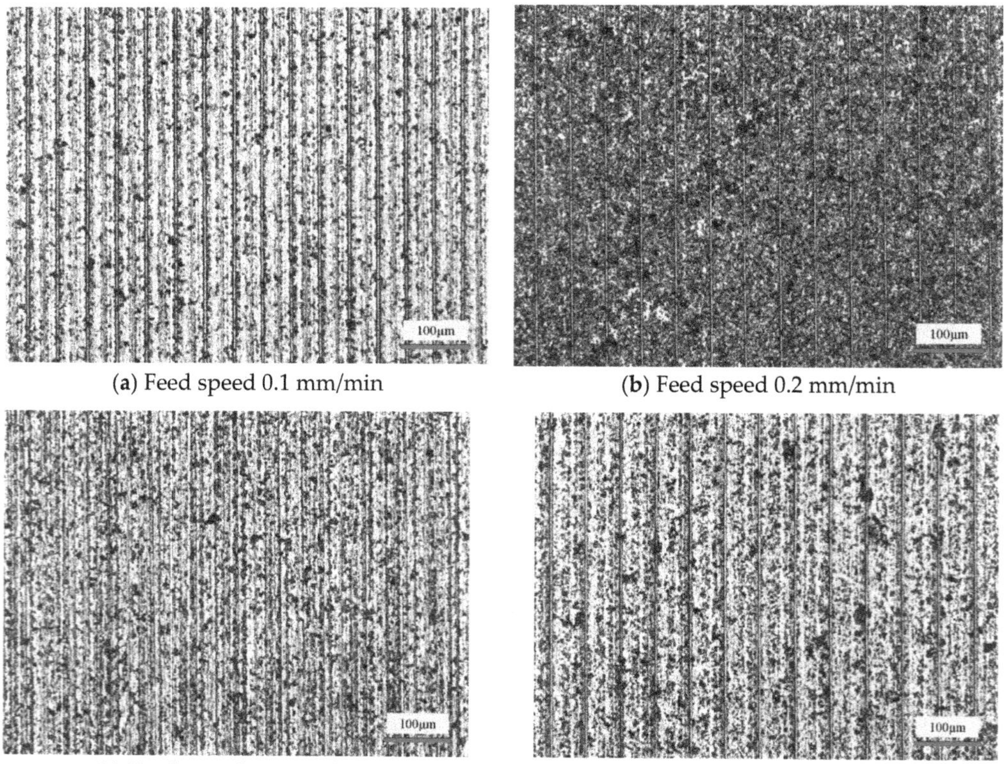

(**a**) Feed speed 0.1 mm/min (**b**) Feed speed 0.2 mm/min

(**c**) Feed speed 0.3 mm/min (**d**) Feed speed 0.4 mm/min

Figure 9. Surface morphology of slices at different feed speeds (wire speed 1200 m/min).

3.2. Surface Roughness R_a and Waviness W_a of the Sliced Surface

Measuring and analyzing the 3D surface images of the Si_3N_4 slices, three curves can be derived as shown in Figure 10. Figure 10a exhibits the overall height variation in the sliced surface in the specimen feed direction. Figure 10b,c show the surface roughness and waviness curves of the sliced surface, respectively. The fluctuation of the roughness curves shown in Figure 10b is fairly uniform. A stable roughness value can be obtained over the whole range of the sliced surface. It shows that the diamond wire saw has good processing stability when cutting Si_3N_4 ceramics. As can be seen from Figure 10c, the waviness curve on the surface of the Si_3N_4 slice shows an irregular curve. The shape of this waviness curve is different from that exhibited when cutting and processing crystalline materials such as silicon crystals and sapphire crystals. According to the results of previous studies, the surface of the cut crystalline material has a sinusoidal-like rippled pattern [8,24]. In contrast, the surface waviness of the Si_3N_4 slices in this experiment is more like a high frequency curve similar to irregular vibration. At the two edges of Figure 10c, the difference between neighboring peaks and troughs is between 0.6 and 0.8 μm. While in the middle region of the curve, the difference between adjacent peaks and troughs is around 0.2–0.3 μm. This phenomenon may be related to the material properties of Si_3N_4 itself. Si_3N_4 is a ceramic-like material made of powder pressed at high temperature and high pressure. The internal microscopic ceramic is prone to uneven density, and even produces local internal stress. The diamond saw wire, as a flexible tool, is easily affected by these factors in the feeding direction. Thus, the microscopic feed trajectory deviates, resulting in this unique surface waviness curve in the Si_3N_4 cutting process.

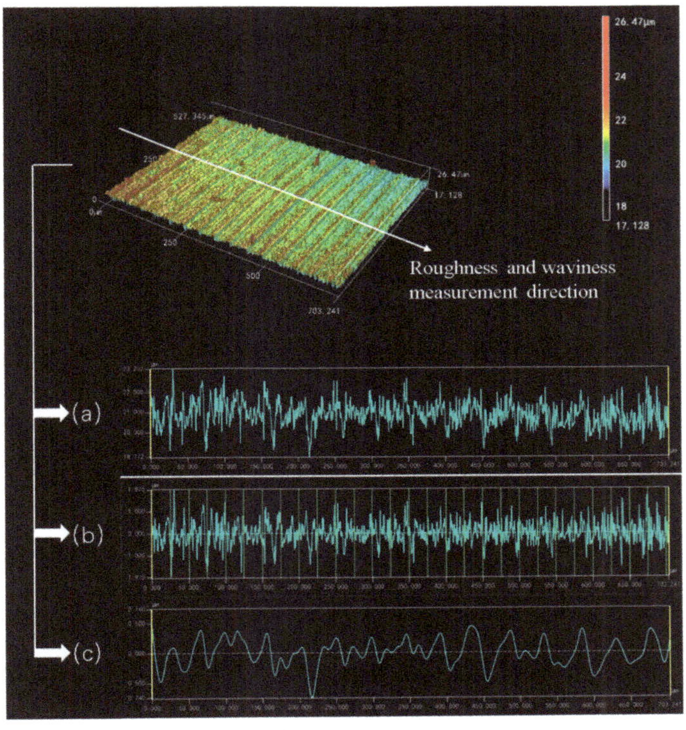

Figure 10. Sliced surface morphology curves; (**a**) measured cross-section graph, (**b**) surface roughness R_a graph, and (**c**) Waviness W_a graph. (Wire speed 1000 m/min and feed speed 0.3 mm/min).

3.2.1. Effect of Wire Speed on Surface Roughness and Waviness of Slices

The three-dimensional surface morphology can reflect the quality of the machined surface very intuitively. From Figure 11, it can be seen that the surfaces of the Si_3N_4 ceramic slices gradually become smooth and flat with the increase in wire speed. Especially from 800 m/mim to 1200 m/min, the variation range of the surface height is 9.005 μm, 8.131 μm and 7.028 μm, which shows that the increase in the wire speed is favorable to improve the surface quality. However, an increase in surface defects occurs locally in the slices during high-speed cutting. Although the cutting ability becomes stronger as the wire speed increases, the vibration amplitude of the wire may also increase. So, under the mutual constraints of two different factors, increasing the wire speed endlessly does not lead to a significant improvement in the surface quality.

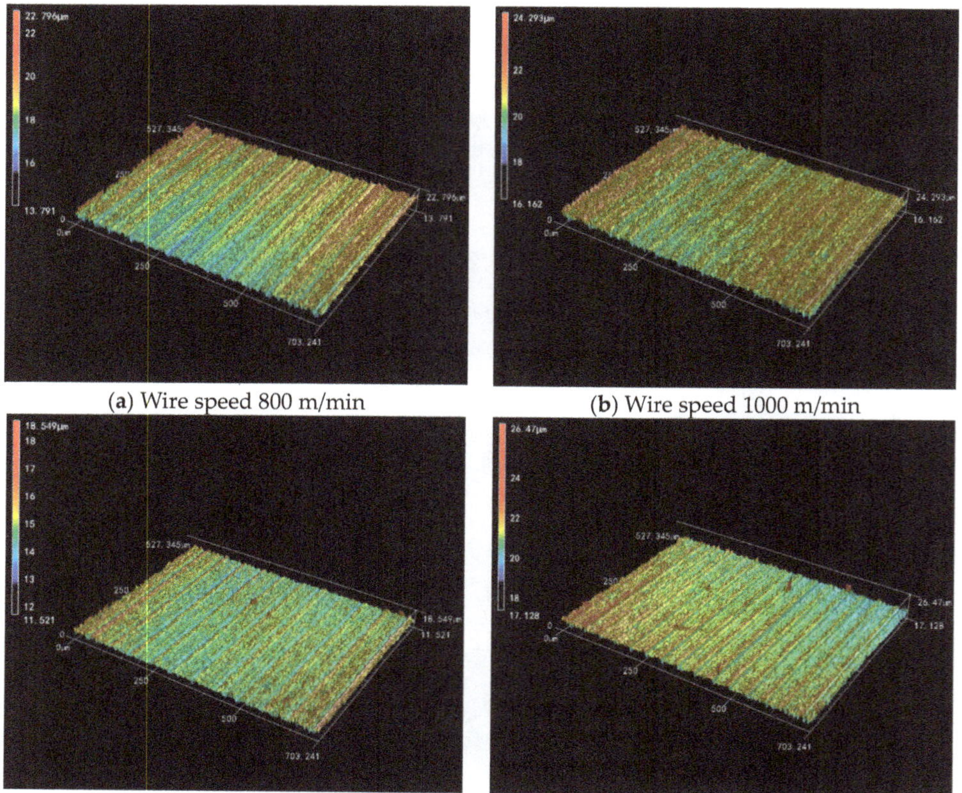

(**a**) Wire speed 800 m/min (**b**) Wire speed 1000 m/min

(**c**) Wire speed 1200 m/min (**d**) Wire speed 1600 m/min

Figure 11. The 3D surface morphology of slices at different wire speeds (feed speed 0.3 mm/min).

Figure 12 shows the effect of wire speed on the surface roughness and waviness of Si_3N_4 slices at a feed speed of 0.3 mm/min. The surface roughness and waviness of Si_3N_4 slices decreased with increasing wire speed. Under four sets of wire speed parameters, the R_a between neighboring groups decreased by 7.75% and 7.25%, as well as 1.88%. Among them, the decrease in roughness between neighboring groups is more obvious when the wire speed is low. As the wire speed is further increased, especially between the high wire speed of 1600 m/min and 1200 m/min, the value of R_a decreases insignificantly and finally stabilizes in the range of 0.31–0.32 μm. It can be seen that under low-speed cutting conditions, the change in wire speed has a direct effect on the change in the R_a value. Under the high-speed cutting condition, the change in wire speed has less effect on R_a. In addition to this, the value of W_a between neighboring groups decreases by 30.35% and 4.29%, as well as 3.73% when the wire speed increases. The variation in waviness with wire speed also has a similar pattern. However, in the overall curve, it seems that the change in wire speed does not affect the roughness and waviness to a large extent, and the differences between groups are small. Therefore, when processing Si_3N_4 ceramics, it is not possible to pursue a too high wire speed, which places higher demands on equipment performance.

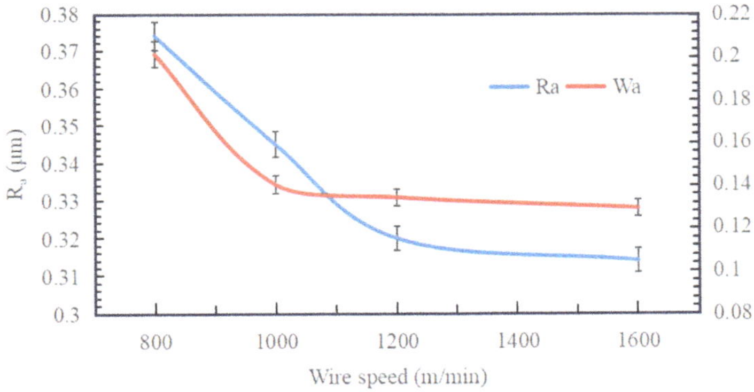

Figure 12. Effect of wire speed on surface roughness and waviness of slices (feed speed 0.3 mm/min).

3.2.2. Effect of Feed Speed on Surface Roughness and Waviness of Slices

Figure 13 shows the 3D surface morphology images derived from Experiment 2, it can be seen that the height of the Si_3N_4 surface keeps changing with the increase in the feed speed. The variation ranges from 6.511 μm, 6.947 μm and 7.028 μm and finally increases to 7.07 μm. It can be learned that the range of height undulation increases with the increase in feed speed. The two have a positive correlation. Compared with the 3D surface morphology images of Experiment 1, the increase in feed speed is more inclined to a uniform decrease in the overall surface quality.

Figure 14 shows the effect of the workpiece feed speed on the surface roughness and waviness of Si_3N_4 slices at a wire speed of 1200 m/min. With the increase in feed speed, the R_a and W_a of the Si_3N_4 slices along the feed direction are increasing. Numerically, R_a between neighboring feed speed control groups increased by 11.65% and 3.44%, as well as 5.04%. The change in R_a between neighboring groups is proportional to the increase in the feed speed. The variation in R_a is higher at feed speed values from 0.1 mm/min to 0.2 mm/min. So, the sensitivity of R_a change is higher at a lower feed speed. When the feed speed is further increased, the sensitivity of the change in the value of R_a tends to stabilize and increases steadily with the feed speed. From the W_a curve, it is known that the value of W_a increases by 10.19%, 19.4% and 6.94% between neighboring groups when the feed speed increases. It can be seen that the variation in W_a is also positively related to the increase in feed speed and varies more in the range of 0.2–0.3 mm/min. In addition to this, the variation in W_a is also similar to the variation in R_a.

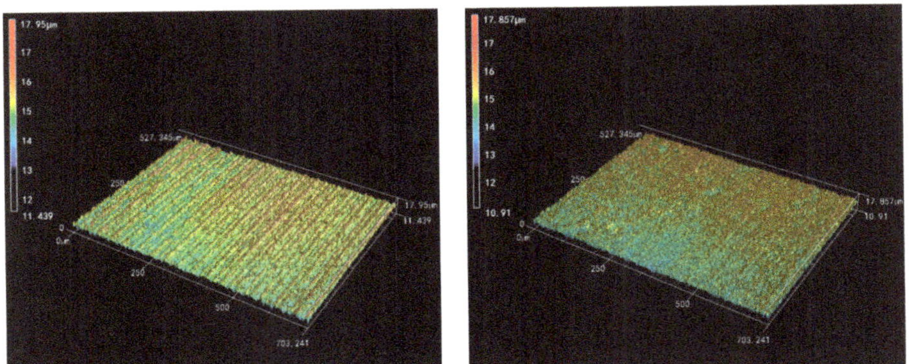

(**a**) Feed speed 0.1 mm/min (**b**) Feed speed 0.2 mm/min

Figure 13. *Cont.*

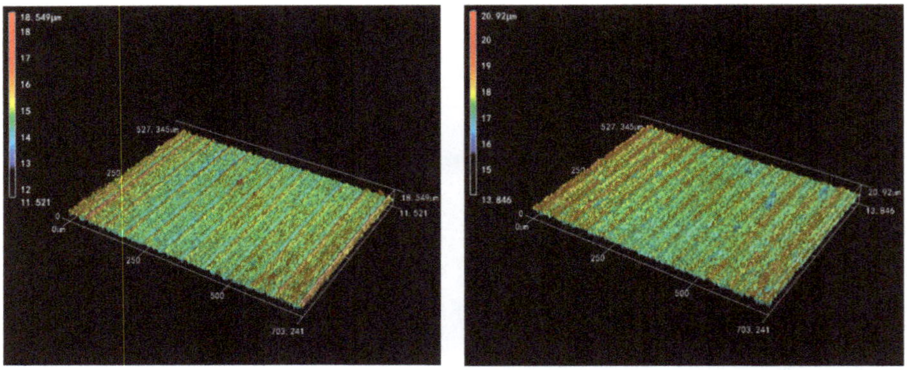

(c) Feed speed 0.3 mm/min (d) Feed speed 0.4 mm/min

Figure 13. The 3D surface morphology of slices at different feed speeds (wire speed 1200 m/min).

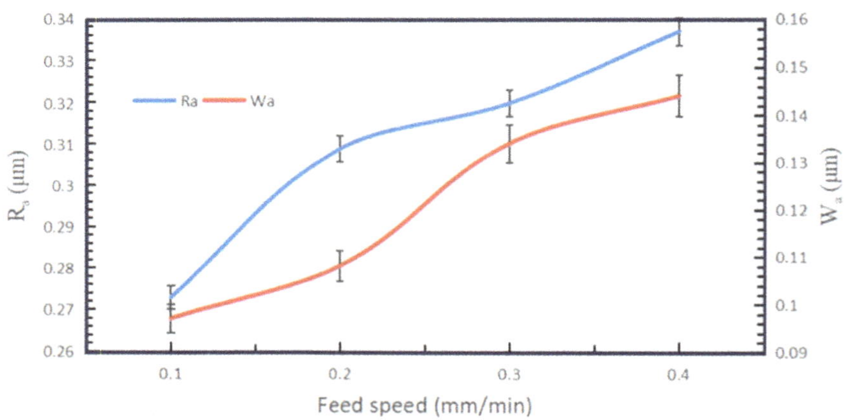

Figure 14. Effect of feed speed on surface roughness and waviness of slices (wire speed 1200 m/min).

4. Conclusions

This paper carries out an industrial diamond wire saw slicing experiment on Si_3N_4 ceramics. The influence law of sawing process parameters used in industrial practical application on the surface morphology, surface roughness and waviness of Si_3N_4 ceramic slices were analyzed. The following conclusions are obtained.

Under the sawing processing conditions for industrial applications, the material removal of Si_3N_4 ceramics surfaces in diamond wire sawing is predominantly in a brittle mode. The slice morphology shows brittle pits and regularly distributed wire marks in the 20–55 μm scale range.

The surface roughness along the feed direction of the workpiece varied in the range of 0.27–0.38 μm and the surface waviness varied in the range of 0.09–0.21 μm. These two indicators decreased with the increase in wire speed and the decrease in feed speed. The trends of surface roughness and waviness have a good correspondence with the surface morphology of the slices.

Author Contributions: S.Z.: methodology, writing—original draft preparation, data curation and visualization. Y.G. (Yufei Gao): conceptualization, methodology, writing—review and editing and project administration. X.Z.: conceptualization and methodology. Y.G. (Yufeng Guo): conceptualization. All authors have read and agreed to the published version of the manuscript.

Funding: The work is financially supported by the Natural Science Foundation of Shandong Province (No. ZR2023ME145), National Natural Science Foundation of China (No.51875322) and National Key Research and Development Project (Nos. 2022YFB3401901, 2022YFB3401902).

Data Availability Statement: All data generated or analyzed during this study are included in this published article.

Conflicts of Interest: The authors declare no conflict of interest.

References

1. Lv, X.; Huang, J.; Dong, X.; Yan, Q.; Ge, C. Influence of alpha-Si_3N_4 coarse powder on densification, microstructure, mechanical properties, and thermal behavior of silicon nitride ceramics. *Ceram. Int.* **2023**, *49*, 21815–21824. [CrossRef]
2. Mir, A.H.; Ahmad, S.N. A study on fabrication of silicon nitride-based advanced ceramic composite materials via spark plasma sintering. *Proc. Inst. Mech. Eng. Part L J. Mater.* **2021**, *235*, 1739–1756.
3. Hu, F.; Zhu, T.; Xie, Z.; Liu, J. Effect of composite sintering additives containing non-oxide on mechanical, thermal and dielectric properties of silicon nitride ceramics substrate. *Ceram. Int.* **2021**, *47*, 13635–13643. [CrossRef]
4. Pramanick, A.; Mandal, S.; Dey, P.P.; Das, P.K. WEDM process optimization of sintered structural ceramic sample by using fuzzy-MPCI technique. *Mater. Today Proc.* **2021**, *41*, 925–934. [CrossRef]
5. Bharathi, V.; Anilchandra, A.R.; Shantanu, S.S. A review on the challenges in machining of ceramics. *Mater. Today Proc.* **2021**, *46*, 1451–1458.
6. Weixler, J.; Zweifel, M.; Schudeleit, T.; Bambach, M.; Wegener, K. Laser ablation study of cutting ceramics with consideration of the beam inclination angle. *Materials* **2023**, *16*, 2509. [CrossRef]
7. Srinivasana, V.P.; Palanib, P.K. Surface integrity, fatigue performance and dry sliding wear behaviour of Si_3N_4–TiN after wire-electro discharge machining. *Ceram. Int.* **2020**, *46*, 10734–10739. [CrossRef]
8. Li, X.; Gao, Y.; Yin, Y.; Wang, L.; Pu, T. Experiment and theoretical prediction for surface roughness of PV polycrystalline silicon wafer in electroplated diamond wire sawing. *J. Manuf. Process.* **2020**, *49*, 82–93. [CrossRef]
9. Liang, L.; Li, S.; Lan, K.; Yu, R.; Wang, J.; Zhao, W. Experimental study on the influence of wire-saw wear on cutting force and silicon wafer surface. *Micromachines* **2023**, *16*, 3619. [CrossRef]
10. Gao, Y.; Chen, Y. Sawing stress of sic single crystal with void defect in diamond wire saw slicing. *Int. J. Adv. Manuf. Technol.* **2019**, *103*, 1019–1031.
11. Zhu, Z.; Gao, Y.; Zhang, X. Study on subsurface microcrack damage depth of diamond wire as-sawn sapphire crystal wafers. *Eng. Fract. Mech.* **2023**, *28*, 109347. [CrossRef]
12. Wang, N.; Jiang, F.; Xu, X. Research on the machinability of A-plane sapphire under diamond wire sawing in different sawing directions. *Ceram. Int.* **2019**, *45*, 10310–10320. [CrossRef]
13. Qiu, J. Fundamental research on machining performance of diamond wire sawing and diamond wire electrical discharge sawing quartz glass. *Ceram. Int.* **2023**, *48*, 24332–24345. [CrossRef]
14. Ban, X.; Li, Y.; Han, S.; Qiu, H.; Wang, X.; Cui, Z. Parameters optimization for ferrite slicing based on grey theory. *Diam. Abras. Eng.* **2022**, *42*, 332–337.
15. Yang, C.; Wang, Z.; Su, H.; Fu, Y.; Zhang, N.; Ding, W. Numerical analysis and experimental validation of surface roughness and morphology in honing of Inconel 718 nickel-based superalloy. *Adv. Manuf.* **2023**, *11*, 130–142. [CrossRef]
16. Yin, Y.; Gao, Y. Experimental study on slicing photovoltaic polycrystalline silicon with diamond wire saw. *Mat. Sci. Semicon. Proc.* **2020**, *106*, 104779–104789. [CrossRef]
17. Liu, Y.; Gao, Y.; Yang, C. Analysis of sawing characteristics of fine diamond wire slicing multicrystalline silicon. *Diam. Relat. Mater.* **2021**, *12*, 108708. [CrossRef]
18. Qiu, J.; Li, X. Surface formation, morphology, integrity and wire marks in diamond wire slicing of mono-crystalline silicon in the photovoltaic industry. *Wear* **2022**, *488–489*, 204186. [CrossRef]
19. Qiu, J.; Li, X. Formation mechanism of wire bow and its influence on diamond wire saw process and wire cutting capability. *Int. J. Mech. Sci.* **2020**, *185*, 105851–105862. [CrossRef]
20. Costa, E.C.; Santos, C.P.; Carvalho, V.A.; Xavier, F.A. Study on surface integrity and ductile cutting of PV polycrystalline silicon and wear mechanisms of electroplated diamond wire. *Int. J. Adv. Manuf. Technol.* **2022**, *122*, 1539–1553. [CrossRef]
21. Huang, H.; Zhang, Y. Experimental investigation on the machining characteristics of single-crystal SiC sawing with the fixed diamond wire. *Int. J. Adv. Manuf. Technol.* **2015**, *81*, 955–965. [CrossRef]
22. Liang, L.; Li, S.; Lan, K.; Wang, J.; Yu, R. Fixed-diamond abrasive wire-saw cutting force modeling based on changes in contact arc lengths. *Micromachines* **2023**, *14*, 1275. [CrossRef] [PubMed]
23. Yang, C.; Su, H.; Gao, S.; Ai, Q.; Fu, Y.; Ding, W.; Xu, J. Characterization and life prediction of single-pass honing tool for fuel injection nozzle. *Chin. J. Aeronaut.* **2021**, *34*, 225–240. [CrossRef]
24. Zhu, Z.; Gao, Y.; Shi, Z.; Zhang, X. Study on surface characteristics of as-sawn sapphire crystal wafer considering diamond saw wire wear. *Wear* **2023**, *530–531*, 205037. [CrossRef]

25. Liu, Q.; Cheng, J.; Liao, Z.; Luo, X.; Yang, Y.; Li, M.; Yang, H.; Tan, C.; Wang, G.; Ding, W.; et al. Research on the light intensity modulation and characterizing methods of surface texture on KDP optics generated in fly-cutting and micro ball-end milling processes. *CIRP J. Manuf. Sci. Technol.* **2023**, *41*, 30–43. [CrossRef]
26. Liu, Q.; Cheng, J.; Liao, Z.; Liu, M.; Chen, M.; Zhao, J.; Lei, H.; Ding, W. Fractal analysis on machined surface morphologies of soft-brittle kdp crystals processed by micro ball-end milling. *Materials* **2023**, *16*, 1782. [CrossRef]
27. Huang, S.; Gao, S.; Huang, C.; Huang, H. Nanoscale removal mechanisms in abrasive machining of brittle solids. *Diam. Abras. Eng.* **2022**, *42*, 257–267.
28. Li, C.; Piao, Y.; Zhang, F.; Zhang, Y.; Hu, Y.; Wang, Y. Understand anisotropy dependence of damage evolution and material removal during nanoscratch of MgF_2 single crystals. *Int. J. Extrem. Manuf.* **2023**, *5*, 015101. [CrossRef]

Disclaimer/Publisher's Note: The statements, opinions and data contained in all publications are solely those of the individual author(s) and contributor(s) and not of MDPI and/or the editor(s). MDPI and/or the editor(s) disclaim responsibility for any injury to people or property resulting from any ideas, methods, instructions or products referred to in the content.

micromachines

MDPI

Article

Longitudinal–Torsional Ultrasonic Grinding of GCr15: Development of Longitudinal–Torsional Ultrasonic System and Prediction of Surface Topography

Huan Zhang [1], Ying Niu [1,*], Xiaofeng Jia [2,*], Shuaizhen Chu [1] and Jingjing Niu [1]

1 School of Mechanical and Power Engineering, Henan Polytechnic University, Jiaozuo 454000, China
2 School of Mechanical Engineering, Anyang Institute of Technology, Anyang 455000, China
* Correspondence: niuying@hpu.edu.cn (Y.N.); jiaxf@ayit.edu.cn (X.J.)

Abstract: The common material of bearing rings is GCr15 bearing steel which is a typical difficult-to-machine material. As an important working surface of the bearing, the inner surface of the raceway plays a vital role in the performance of the bearing. As an important means to solve the high-performance manufacturing of difficult-to-machine materials, longitudinal–torsional ultrasonic processing is widely used in various types of processing. In the presented work, the basic size of the horn is obtained from the wave equation of the forced vibration, and the modal analysis and amplitude test are carried out to verify the rationality of the LUTG structure. Then, according to the probability density function of cutting thickness and the overlapping effect of adjacent abrasive trajectories, the LUTG surface topography prediction model is established by using the height formula of the surface residual material, and the model reliability is verified by using the orthogonal test. The error between the test results and the prediction model is within 13.2%. Finally, based on the response surface method, the optimal process parameters that can meet the requirements of low roughness (Ra) and high material removal rate (MRR) are screened, and the optimal combination of process parameters is obtained as follows: A = 4.5 μm, n = 6493.3 r/min, a_p = 28.4 μm, and v_f = 21.1 mm/min.

Keywords: GCr15; device; surface topography prediction; response surface

1. Introduction

Rolling bearings have the ability to withstand radial loads and single-direction longitudinal loads in rotating mechanisms and are widely used in the aviation and automotive industries and other fields [1–3]. The bearing ring is the key component of the rolling bearing. The surface of the bearing ring's internal circle is an important working surface and plays a vital role in the bearing ring's performance [4,5]. As a high performance bearing material, Gcr15 bearing steel has high hardness, high strength, and high wear resistance. It is often used in bearing parts such as sleeves and rolling elements. However, GCr15 bearing steel is a typical difficult-to-machine, hard, and brittle material. During processing, if defects such as burns and cracks occur on the surface of the sleeve raceway, they can cause early fatigue and internal surface peeling of the bearing [6–8].

Ultrasonic vibration processing technology can reduce the force and heat during processing and improve the surface quality of workpieces. It has shown outstanding advantages in aerospace, the automobile industry, and other technical fields and has become one of the important means to solve the high performance manufacturing of difficult-to-machine materials.

Compared with ordinary processing technology, ultrasonic processing can reduce force and heat during processing and improve the surface quality of the workpiece. Zheng et al. [9] studied ultrasonic vibration grinding of zirconia, established a grinding temperature model, and compared ultrasonic grinding with ordinary grinding tests

The results show that ultrasonic vibration can reduce grinding temperature. Xu et al. [10] studied the ultrasonic vibration milling of titanium alloy Tc4 and aluminum alloy 606t6. The test results show that compared with traditional milling, ultrasonic vibration milling can effectively reduce the milling force, reduce the surface defects of the workpiece, and prolong the service life of the tool.

Compared with one-dimensional ultrasonic processing technology, longitudinal–torsional ultrasonic processing technology can further reduce force and heat during processing and improve surface quality [11,12]. Niu et al. [13] used longitudinal–torsional ultrasonic milling to process the difficult-to-machine material titanium alloy and compared the cutting force under ordinary ultrasonic milling and longitudinal–torsional ultrasonic milling via experiments. The results show that the cutting force under longitudinal–torsional ultrasonic milling is lower. Chen et al. [14] studied the longitudinal–torsional and ordinary ultrasonic vibration grinding of silicon carbide ceramics. Under the same grinding conditions, LTUG has a smaller grinding force and lower surface roughness than LUAG.

The surface morphology of the parts has a great influence on the contact state, surface wear, lubrication state, friction, and vibration of the parts. Accurately predicting the surface morphology of the machined surface is of great significance for improving the parts' quality. Based on the characteristics of ultrasonic vibration processing, Yang et al. [15] proposed a contact velocity model between abrasive particles and workpieces in ultrasonic vibration grinding combined with a single abrasive particle for adjacent abrasive particles. Chen et al. [16] proposed a surface topography modeling and prediction method for ultrasonic grinding considering ploughing. Assuming that the gravel is spherical, a grinding wheel surface model that considered the random distribution of grains was established. Based on the geometric mapping relationship between grains and workpieces in ultrasonic grinding, a grain cutting model considering real-time cutting depth and ploughing action was proposed. Gao et al. [17] designed a new type of ultrasonic vibration grinding device. Assuming that the abrasive grain is a rigid sphere and considering the relationship between the angle of the grinding trajectory and the axis of the workpiece, the radius of the grinding trajectory and the distance between the abrasive grains, a surface topography theoretical model of ultrasonic vibration grinding was established.

The prediction of grinding surface morphology is mostly based on the kinematics and distribution characteristics of abrasive particles, material removal mechanisms, mathematical algorithms, and so on. However, there are few studies on the influence of elastic deformation of the grinding area on surface morphology.

Based on this, this paper proposes a LTUG process for the inner surface of GCr15 bushing. The influence of elastic deformation between the bushing and the grinding wheel on the length of the abrasive cutting path under ultrasonic action is introduced. Based on the longitudinal–torsional ultrasonic vibration, the maximum undeformed chip thickness formula under the action of multiple abrasive grains affected by elastic deformation is established by using the cutting thickness probability density function. The LTUG morphology prediction model is established according to the height value of the surface residual material, and the model is verified. Finally, the surface roughness value is calculated by using the established surface morphology model. The effects of different process parameters and processing conditions on surface roughness and surface morphology are studied.

2. Development of LTUG System

In the presented work, the grinding wheel is selected as the carrier of vibration. The structure diagram of the ultrasonic vibration grinding system is shown in Figure 1. It is composed of an ultrasonic power supply, transducer, horn, sleeve, wireless transmission unit, and grinding wheel. The ultrasonic power supply converts the alternating current into an ultrasonic frequency electrical signal and transmits the electrical signal to the wireless transmission system. The wireless transmission system transmits the electrical signal to the transducer, and the transducer converts the ultrasonic frequency electrical signal into a sinusoidal mechanical vibration and transmits it to the horn. The horn amplifies the mechanical vibration output by the transducer.

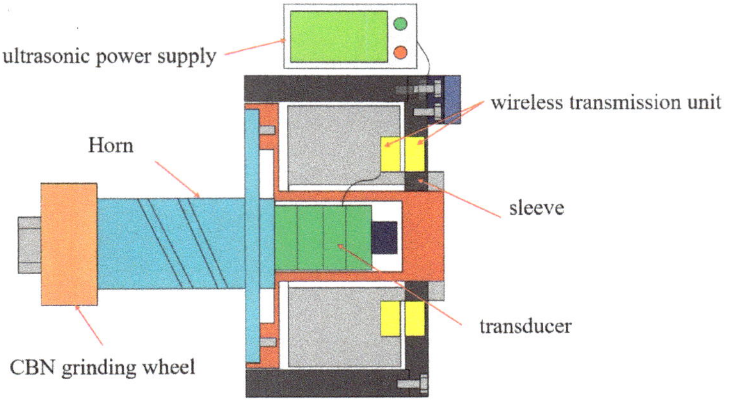

Figure 1. Schematic diagram of the LTUG system.

2.1. Development of Longitudinal Horn

As shown in Figure 2a, the axial direction of the rod is taken as the x-axis, and the infinitesimal segment dx is taken at x on the rod. The axial displacement at the left end is $u(x)$, while the axial displacement at $x + dx$ on the rod is $u + \frac{\partial u}{\partial x}dx$. The deformation in the dx segment is $\frac{\partial u}{\partial x}dx$, while ε represents the strain, and σ represents the stress [18].

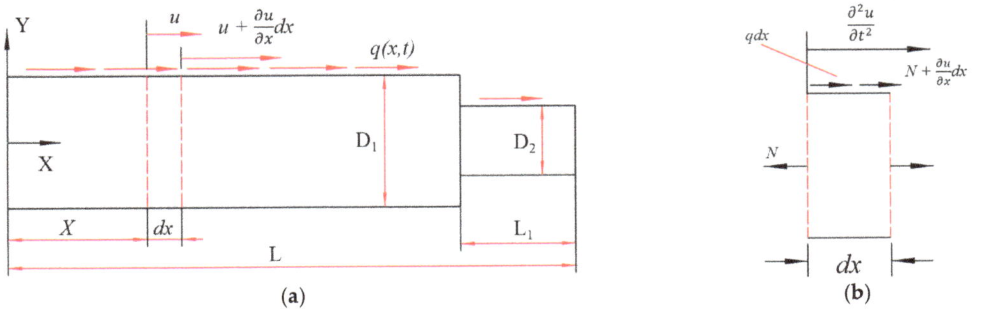

Figure 2. Longitudinal vibration of the horn. (**a**) Longitudinal vibration of an equal straight horn. (**b**) Force analysis diagram of the micro segment.

The force analysis diagram of the infinitesimal segment dx is shown in Figure 2b. According to Newton's second law [19]:

$$\rho A dx \frac{\partial^2 u}{\partial t^2} = \left(N + \frac{\partial N}{\partial x}dx\right) - N + q(x,t)dx \quad (1)$$

The above formulas are combined to obtain [20].

$$\frac{\partial^2 u}{\partial t^2} = \left(\frac{E}{\rho}\right)\frac{\partial^2 u}{\partial x^2} + \frac{1}{\rho A}q(x,t) \quad (2)$$

Formula (4) is the wave equation of the forced vibration of the rod. In the formula, $\frac{E}{\rho} = a^2$ represents the longitudinal propagation speed of the elastic wave along the rod. The energy exists in the form of a wave during the transmission process. The full wavelength is

$\lambda = \frac{a}{f}$ where f represents the design frequency. The transmission form of the wave in the rod is a sine curve, and if the horn is too long, it will lead to bending deformation due to insufficient stiffness of the rod; if it is too short, it cannot guarantee that the spindle can completely process the inner circle of the sleeve. Therefore, a wavelength is selected as the boundary condition of the wave equation. Because of the good processing performance and vibration energy transmission effect of 45 steel, 45 steel is selected as the horn material. The design frequency of the horn is 28 kHz.

According to the above formula and analysis, the size of the horn is obtained as shown in Table 1, and the model is imported into the analysis software ANSYS 2021R1. The modal analysis of the model is carried out to verify the rationality of the geometric structure of the horn.

Table 1. Structural parameters of the longitudinal amplitude horn.

Type	L/mm	L1/mm	D1/mm	D2/mm
Parameter	182	33	38	20

It can be seen from Figure 3 that the vibration effect at the flange is the smallest, and the vibration effect at the small end of the grinding wheel is the largest, showing a better vibration mode. This is because during the transmission of vibration energy in the horn, the small end is small in size, and the energy is concentrated. It can be seen from the above that the horn presents a longitudinal vibration mode, and the deviation between the natural frequency and the design frequency is small, so the structural design of the horn is more reasonable.

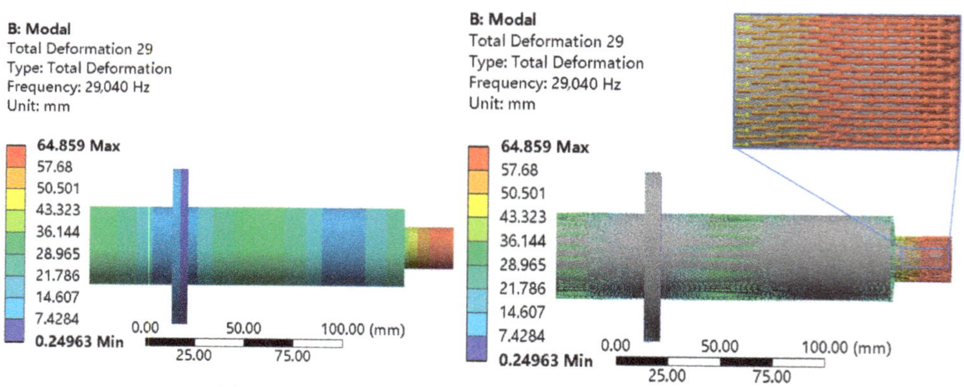

Figure 3. Modal analysis of the longitudinal horn. (**a**) Total deformation cloud image. (**b**) Total deformation vector diagram.

2.2. Design of Longitudinal–Torsional Ultrasonic Horn

Adding a spiral groove with a certain geometric structure on the horn can produce a certain amplitude of circumferential vibration. Four spiral grooves with a spiral angle of 45°, a width of 6 mm, and a vertical length of 90 mm are added. The specific simulation results are shown in Figure 4. It can be known from the vibration mode vector that the vibration vector direction of the horn with four spiral grooves corresponds to the change in direction of the spiral groove, resulting in obvious torsional vibration.

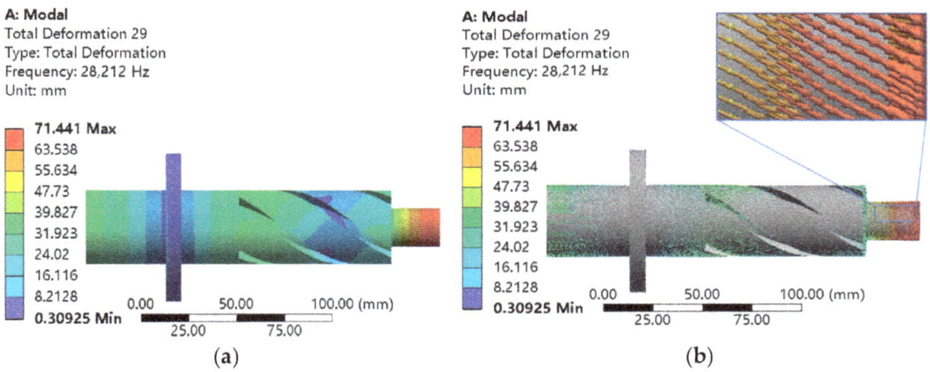

Figure 4. Modal analysis of the longitudinal–torsion horn. (**a**) Total deformation cloud image. (**b**) Total deformation vector diagram.

After obtaining the horn with the determined geometric structure, the models of the longitudinal–torsional ultrasonic horn and the grinding wheel are assembled to obtain the 3D model of the longitudinal–torsional ultrasonic horn. In order to prevent excessive consumption of ultrasonic vibration energy in the working process, the aluminum alloy with a lighter weight is used as the grinding wheel matrix. The model is imported into the analysis software ANSYS, and the modal simulation results are shown in Figure 5.

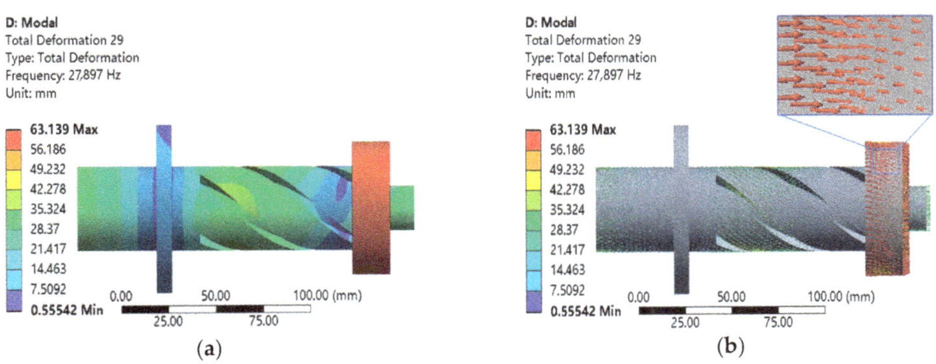

Figure 5. Modal analysis of the LTUG device. (**a**) Total deformation cloud image. (**b**) Total deformation vector diagram.

From Figure 5, it can be seen that the resonant frequency of the ultrasonic amplitude transformer is 27,897 Hz, which is very close to the design frequency. The vibration at the flange of the longitudinal torsional ultrasonic vibration grinding system is the weakest, and the vibration at the front end of the grinding wheel is the strongest, which is in line with the expected effect.

2.3. Amplitude Test of LTUG System

The amplitude of the LTUG system was measured using a laser displacement sensor. The ultrasonic amplitude measurement site is shown in Figure 6. Four places were evenly selected in the direction of the end face and the circumferential direction of the grinding wheel to measure the longitudinal and torsional vibration amplitudes in both directions. The average value was used as the vibration amplitude of the longitudinal–torsional ultrasonic vibration grinding device. After measurement, it is known that the torsional vibration amplitude at the same frequency is about 25% of the longitudinal amplitude (after averaging the measured data, as shown in Figure 7). After analyzing the amplitude of the LTUG system at the same frequency and with multiple tests of the longitudinal and torsional amplitude of the grinding wheel, it is known that the torsional amplitude of each point in the circumferential direction of the grinding wheel is about 25% of the longitudinal amplitude.

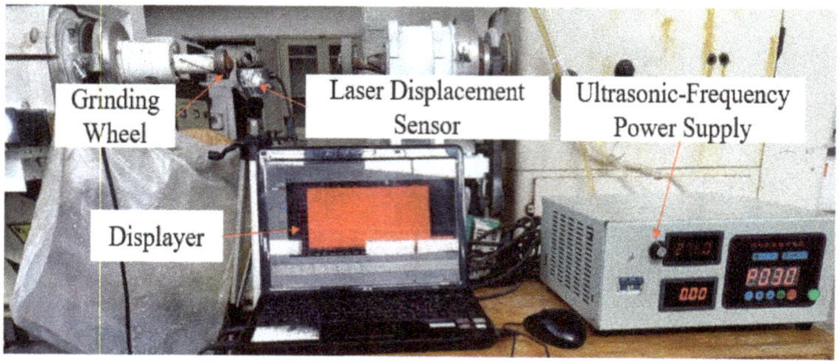

Figure 6. On-site ultrasonic amplitude measurement.

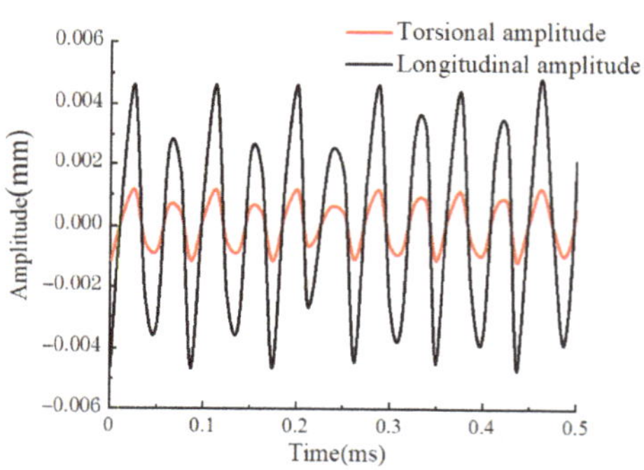

Figure 7. LTUG system amplitude test results.

3. Surface Topography Prediction of LTUG

At present, the prediction of grinding surface morphology is mostly based on the kinematic trajectory of abrasive particles, material removal mechanisms, the distribution characteristics of abrasive particles, and mathematical algorithms [21], and the influence of elastic deformation between workpieces and grinding wheels on the prediction results is not considered. In this paper, the influence of elastic deformation between the shaft sleeve and the grinding wheel on the length of the cutting path of the abrasive particles under the action of longitudinal torsional ultrasound is introduced. Using the probability density function of the cutting thickness, the formula of the maximum undeformed chip thickness under the action of multiple abrasive particles affected by elastic deformation is established. According to the height of the surface residual material, the LTUG morphology prediction model is established.

In the process of grinding, due to the uneven distribution of abrasive particles on the surface of the grinding wheel and the irregular contour of the abrasive particles, the abrasive particles will process the surface of the workpiece by scraping, ploughing, and cutting. In order to facilitate the modeling, it is assumed that the size and distribution of the abrasive particles on the grinding wheel are uniform and the protrusion height of the abrasive particles is the same, ignoring the effects of the abrasive ploughing and the material fracture on the surface morphology and ignoring the wear of the abrasive particles.

3.1. Maximum Undeformed Cutting Thickness Model Considering Elastic Deformation

In the grinding area, the influence of elastic deformation on the arc length of single abrasive grain grinding is mainly divided into two aspects: (1) the additional length l_c' which is caused by the elastic deformation between the abrasive grain and the workpiece and (2) the additional length a' which is caused by the elastic deformation between the grinding wheel and the workpiece. In order to calculate the actual contact arc length affected by the elastic deformation under the action of longitudinal–torsional ultrasonic internal grinding, it is necessary to calculate the additional contact arc length caused by the two factors according to the elastic deformation relationship between the grinding wheel, the abrasive grain, and the workpiece.

The elastic deformation relationship between the abrasive particles and the workpiece is shown in Equation (3), where $K_i = \frac{1-u_i^2}{\pi E_i}$ and $i = g, \zeta$ represent the abrasive particles and the workpiece, respectively. E_i and u_i represent the elastic modulus and Poisson's ratio of the two materials, respectively. F_n is the normal grinding force of a single abrasive particle under ultrasonic grinding.

$$\delta' = \sqrt[3]{(\frac{3\pi}{2\sqrt{2}})^2 (K_\zeta + K_g)^2 (\frac{1}{d_{g_{max}}}) F_n^2} \tag{3}$$

The total additional length caused by elastic deformation is:

$$l_f \cong l_c' + a' \tag{4}$$

where $l_c' \cong \sqrt{2r\delta}$, $a' = \sqrt{5.12(\frac{\pi r}{b})(K_w + K_s)F_n}$.

Assume:

$$\phi = \sqrt[3]{(\frac{3\pi}{2\sqrt{2}})^2 (K_w + K_g)^3 (\frac{1}{d_{g_{max}}})} \tag{5}$$

$$\psi = \sqrt{5.12(\frac{\pi r}{b})(K_w + K_s)} \tag{6}$$

After finishing, the formula of the arc length of the longitudinal–torsional ultrasonic inner circle contact affected by the elastic deformation is obtained:

$$l = (((\frac{2\sqrt{2\phi r}}{\sqrt[3]{bl_f C}}) F_n^{\frac{1}{3}} + \psi F_n^{\frac{1}{2}})^2 + l_c)^{0.5} \tag{7}$$

The maximum undeformed grinding thickness affected by elastic deformation is:

$$E(h_1) = \sqrt{\frac{(a_p + B)v_f}{2Cl(v_s + v_g)}} \tag{8}$$

Compared with OG, the trajectories of adjacent abrasive particles in the LTUG grinding area are superimposed. Therefore, the superposition of adjacent abrasive trajectories should be considered when analyzing the maximum undeformed cutting thickness of the LTUG. According to the characteristics of LTUG wear particle motion, the function θ is introduced to define the superposition effect of adjacent wear particle motion trajectories. The expression is shown in Equation (9), where ξ_1, ξ_2, ξ_3, ξ_4, and ξ_5 are the indexes that evaluate the correlation between ϑ and its variables. ϑ is the correlation coefficient between ϑ and its variables.

$$\vartheta = \partial \cdot \frac{v_s^{\xi_4} \cdot v_g^{\xi_5}}{a_p{}^{\xi_1} B^{\xi_2} A^{\xi_3}} \tag{9}$$

When combining Equations (8) and (9), according to the superposition effect of adjacent abrasive trajectories in the LTUG process, the expected value of the maximum undeformed cutting thickness under the action of multiple abrasive grains is expressed as follows:

$$E(h') = \vartheta E(h_1) = \partial \cdot \frac{v_s^{\xi_4} \cdot v_g^{\xi_5}}{f^{\xi_1} B^{\xi_2} A^{\xi_3}} \sqrt{\frac{(a_p + B) v_g \left\{ d g_{max}^2 \left(\frac{4\pi}{3\xi} \right)^{\frac{2}{3}} \right\}}{8wl(v_s + v_g)}} \tag{10}$$

The expected value of the maximum undeformed cutting thickness under the action of multiple abrasive grains takes into account many influencing factors, including the influence of the elastic deformation between the abrasive grains, the grinding wheel and the workpiece under the LTUG on the trajectory of the abrasive grains, and the maximum undeformed cutting thickness. In the whole grinding process, there are many abrasive grains involved in grinding. It is necessary to consider the superposition effect of the trajectory of adjacent abrasive grains into the expected value of the maximum undeformed cutting thickness under the action of multiple abrasive grains.

3.2. Surface Topography Model of LTUG

In surface prediction model of LTUG, the inner circle is equivalent to a plane at the microscale. The machined surface of high-precision parts is formed under the combined actions of a large number of abrasive particles. These irregularly shaped and different sizes of abrasive particles are irregularly distributed on the surface of the abrasive tool, resulting in the complexity of the material removal mechanism. When the abrasive particles contact the workpiece surface, the abrasive particles act on the grinding area through sliding friction, ploughing, and cutting stages. Figure 8 describes the formation process of the grinding surface under the action of a large number of random abrasive grains from a microscopic point of view. The surface R is the unprocessed surface, and R^* is the height of a limited number of points on the unprocessed surface. R decreases when the abrasive grains pass through the grinding contact area, forming the grinding surface R^*.

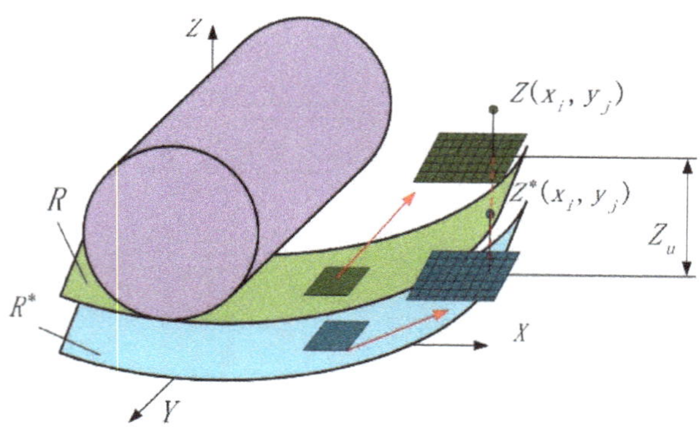

Figure 8. Formation process of the ground surface under the action of random abrasive grains.

The height of any point on the unprocessed surface of the workpiece and the average height of the unprocessed surface are shown in Figure 9. In order to describe the unprocessed surface before grinding, Z_m is defined as the average height of the workpiece surface from the XOY surface before grinding, d_{w-max} is defined as the maximum height of the workpiece surface from the XOY surface before grinding, and d_{w-min} is defined as the minimum height of the workpiece surface from the XOY surface before grinding. $Z_b(x_i, y_j)$ can be used to describe the height value of any random point (x_i, y_j) on the unprocessed surface of the workpiece. According to probability theory, the expression of $Z_b(x_i, y_j)$ is shown in Equation (11), where φ' is the height deviation of the unprocessed surface of the workpiece.

$$Z_b(x_i, y_j) = Z_m + \varphi' \quad \varphi' \in \left[-\frac{d_{w-max} - d_{w-min}}{2}, \frac{d_{w-max} - d_{w-min}}{2} \right] \tag{11}$$

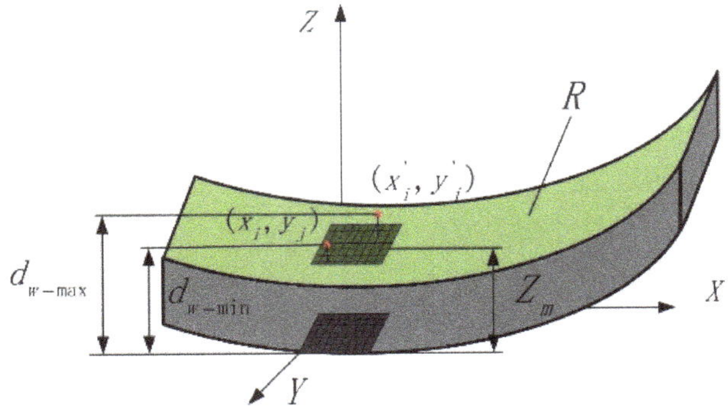

Figure 9. Height and average height of any point on the raw machined surface.

According to the superposition of adjacent abrasive grinding trajectories during LTUG, $Z_u(x_i, y_j)$ is proposed in this study, which means the degree of decline at any point $Z_b(x_i, y_j)$ on the unprocessed surface of the workpiece under the action of random abrasive $G(x_i, y_j)$. The expression is shown in Equation (12), where l_g is the length of the grinding contact zone in the direction of the grinding wheel feed speed, N_{EV} is the number of abrasive grains per unit of grinding wheel volume, a_p is the grinding depth, and v_f is the feed speed.

$$Z_u(x_i, y_j) = \frac{a_p v_f}{\vartheta E(h') N_{EV} b l_g (v_s + v_g)} \tag{12}$$

The expression of N_{EV} is shown in Formula (13). V_t is based on the percentage of the abrasive volume of the grinding wheel, and $\Gamma = d_{g_{max}} - d_{g_{min}}$, $d_{g_{min}}$ is the minimum diameter of the abrasive. d_{gx} is the diameter of a specific abrasive particle, and the diameter of the abrasive particle obeys the normal distribution.

$$N_{EV} = \frac{3V_t \Gamma \sqrt{2\pi}}{4.4\pi \int_{-\Gamma/2}^{\Gamma/2} d_{gx}^3 exp\left[-\frac{1}{2}\left(\frac{4.4}{\delta/2}x\right)^2 \right] dx} \tag{13}$$

The height of the residual material $Z_a(x_i, y_j)$ on the surface of the LTUG is obtained by combining Formulas (11)–(13); that is, the prediction model of the LTUG surface morphology, as shown in Formula (14).

$$Z_a(x_i, y_j) = Z_b(x_i, y_j) - Z_u(x_i, y_j) = Z_m - \frac{a_p v_f}{\vartheta E(h') N_{EV} b l_g v_s} + \varphi' \tag{14}$$

4. The Surface Morphology Model Test of Gcr15 Bearing Ring Internal Circle in LTUG

4.1. Test Conditions and Measurement Methods

In this test, a CNC internal grinder modified by an ordinary lathe is used. The schematic diagram is shown in Figure 10, which is mainly composed of the CNC system, lathe body, grinding wheel dressing device, and LTUG system. Figure 10a,b shows the LTUG test device. It can be seen that the workpiece is fixed by a special three grab chuck. Figure 10c is the working principle diagram of the ultrasonic structure. The motion of the LTUG system is controlled by the CNC system. When the workpiece is processed, the rotation direction of the motorized spindle is opposite to the rotation direction of the workpiece. The motorized spindle feeds radially towards the workpiece, and the roller can round and sharpen the grinding wheel.

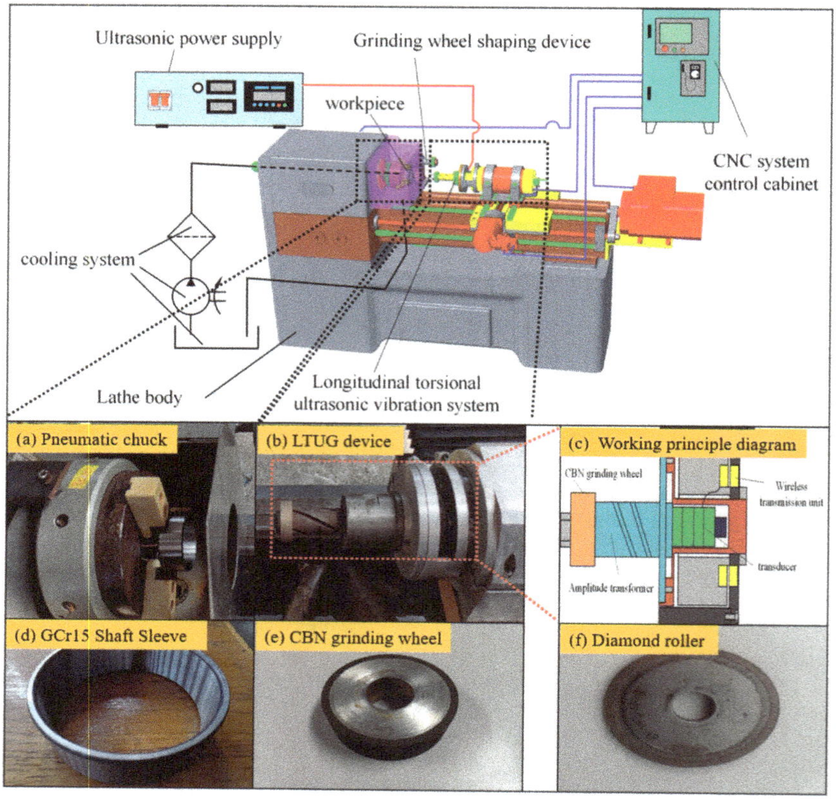

Figure 10. CNC longitudinal torsional ultrasonic internal grinder.

As shown in Figure 10d, the workpiece is a bearing ring with an outer diameter of 110 mm, an inner diameter of 100 mm, and a width of 30 mm. The material is GCr15 bearing steel. Because GCr15 bearing steel has the characteristics of high hardness, high strength, and high wear resistance, the vitrified bond CBN grinding wheel is used in this test. Because of the light weight of aluminum alloy, it is convenient for the vibration energy transfer of the horn. The grinding wheel matrix is selected, and the grain size is 140–170#, as shown in Figure 10e. A metal bond diamond roller is used (as shown in Figure 10f)

After grinding, the workpiece is cut into blocks by an electric spark wire cutting machine. In order to facilitate the observation of the grinding surface morphology and the measurement of its surface roughness, the surface dirt is removed by ultrasonic cleaning with anhydrous ethanol. Then, the surface roughness measuring instrument is used to measure the surface roughness along the grinding's normal direction. The same position of the grinding surface is measured five times, and the average roughness value is taken as the surface roughness value under the group of processing parameters. The three-dimensional structure and linear contour of the workpiece surface are observed by using a laser confocal scanning microscope.

4.2. Test Verification of Surface Topography Prediction Model of LTUG

In order to verify the accuracy of the surface topography prediction model, the workpiece speed is set to 200 r/min, and the LTUG orthogonal test is performed using the grinding parameters shown in Table 2. The LTUG surface topography prediction model Formula (14) is written into a simulation program using MATLAB 2022a software. The data in Table 2 are brought into the surface topography model, and the partial three-dimensional topography simulation results and test results are obtained as shown in Figure 11.

Table 2. Grinding parameters of the orthogonal test.

Number	Revolution Speed n (r/min)	Grinding Depth a_p (um)	Feed Speed v_f (mm/min)	Amplitude A (μm)
1	4000	10	10	0
2	4000	30	30	2
3	4000	50	50	4
4	6000	10	30	2
5	6000	30	50	4
6	6000	50	10	0
7	8000	10	50	4
8	8000	30	10	2
9	8000	50	30	0

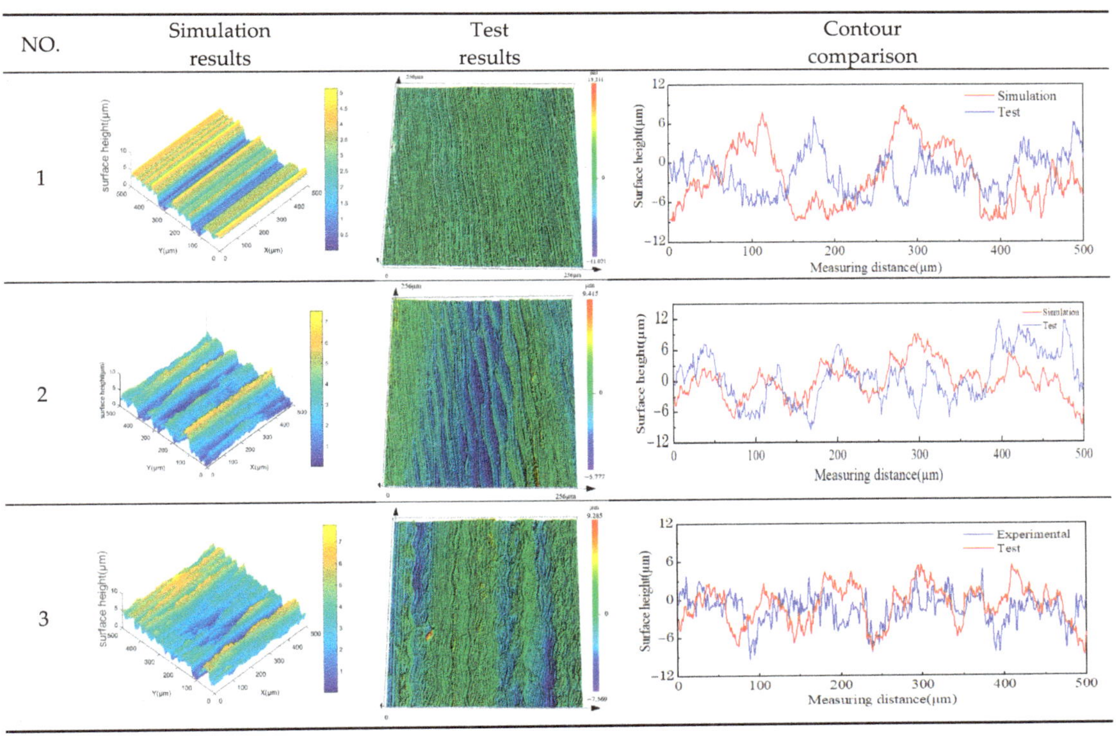

Figure 11. *Cont.*

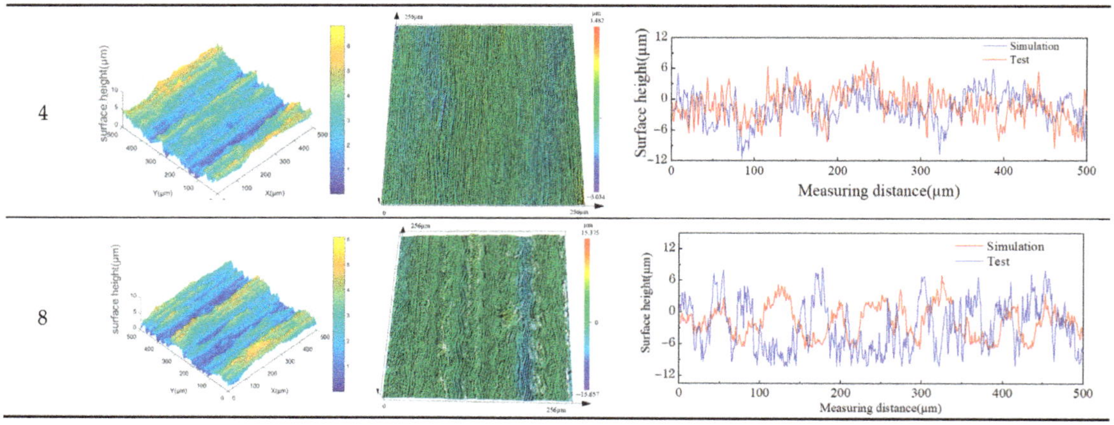

Figure 11. Comparison of simulation and test surface morphology.

The output amplitude of the system is changed by adjusting the frequency and output power of the ultrasonic power supply. The output frequency and current of the ultrasonic power supply fluctuate in a small range to ensure the stability of the output amplitude of the system.

Figure 11 shows the three-dimensional morphology simulation and test diagram under the processing parameters in Table 2 and the three-dimensional morphology diagram taken by the confocal microscope and its normal direction contour diagram. It can be seen from the figure that when $A = 0$, the three-dimensional surface morphology of OG has an obvious straight groove microstructure along the cutting direction. With an appropriate increase in the ultrasonic amplitude, the trajectory of the abrasive particles on the machined surface is superimposed, the distribution of the peak and trough on the machined surface becomes more uniform, and the three-dimensional surface morphology of the LTUG has an obvious periodic sinusoidal microstructure. By comparing the three-dimensional morphology of the two processing methods, it can be found that the three-dimensional fluctuation shape of OG is chaotic and irregular, the three-dimensional morphology fluctuation form of the LTUG rises and falls periodically, and the fluctuation range is lower than that of OG, which is related to the sine wave cutting trajectory of the LTUG abrasive particles. The fluctuation law of the simulation graph and the test measurement graph is basically the same. With the change in the ultrasonic amplitude, the spacing between the peaks of the contour also increases. However, when the ultrasonic amplitude $A = 4$ μm, the surface morphology is significantly different from other grinding morphologies. This is due to the higher vibration energy which leads to larger material shaping protrusions and grooves on the machined surface and increases the surface roughness value.

The two-dimensional contour is extracted from the three-dimensional topography of Figure 11, and the surface roughness value is obtained by using the root mean square method. For each set of data, the actual roughness value is the average of the roughness values of the five grinding areas. The test results are shown in Figure 12. This prediction model can predict the test results, and the prediction accuracy is within 13.2%.

$$\varepsilon\%\ \text{representation error, error} = \frac{\text{test} - \text{simulation}}{\text{test}} \times 100\%$$

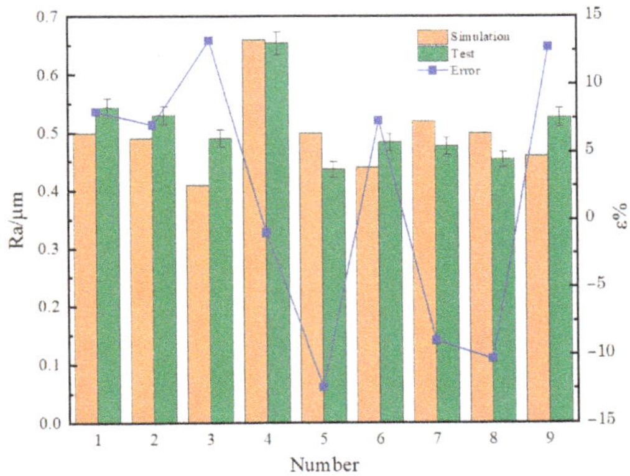

Figure 12. Test and predicted roughness Ra.

4.3. Multi-Objective Parameter Optimization and Verification Based on Response Surface Method

The advantage of the response surface method is that the various levels of the test factors can be continuously analyzed during the optimization of the test strips. According to the response surface diagram, the influence of each variable on the response factors can be visually observed [22]. Under the constraints of the response factors, relevant software can be used to analyze the test parameters to obtain the optimal parameter combination that satisfies the constraints of the response factors. Therefore, the response surface method is used to design experiments that study the effects of parameters and their interactions on surface roughness and material removal rates [23,24].

The test conditions for the material removal rate are as follows: an electronic balance with high precision is selected as the measuring instrument. Before and after each grinding test, anhydrous ethanol ultrasonic cleaning is used to remove impurities and wear debris on the surface. After the water evaporates, the quality of the workpiece is measured. The quality of each workpiece is measured five times, and the average value is taken as the measurement result of the workpiece. The material removal rate is calculated according to the quality of the workpiece before and after grinding. Using the roughness value and the material removal rate as the corresponding factors, the corresponding relationship between the test parameters and the response factors is shown in Table 3.

Table 3. Test plan and response results.

No	a_p	n (r/min)	v_f (mm/min)	A (μm)	Ra (μm)	MRR (mm^3/min)
1	10	4000	10	2	0.43	0.61
2	10	5000	20	3	0.68	0.73
3	10	6000	30	4	0.55	0.84
4	10	7000	40	5	0.51	1.02
5	20	4000	20	4	0.48	0.76
6	20	5000	10	5	0.76	0.93
7	20	6000	40	2	0.64	1.15
8	20	7000	30	3	0.39	1.32
9	30	4000	30	5	0.6	0.97
10	30	5000	40	4	0.43	1.28
11	30	6000	10	3	0.78	1.44
12	30	7000	20	2	0.45	1.68
13	40	4000	40	3	0.5	1.47
14	40	5000	30	2	0.65	1.72
15	40	6000	20	5	0.46	1.82
16	40	7000	10	4	0.4	1.76

Because there are many design variables in this paper and the multivariate quadratic regression equation has 95% accuracy, this paper directly uses the multivariate quadratic regression equation to fit the test data and results of Table 3 and considers all the quadratic terms, primary terms, and interactions between the factors. The curve fitting model of response factor Ra and MRR is obtained using Design-Expert 12 software:

$$\begin{aligned} \text{Ra} = 0.06347 + 0.08778 \times a_p + 0.01342 \times v_f + 1.26333 \times A - 5.84331 \times n \\ +16.73851 \times na_p + 6.66667 \times nA - 0.07333 \times nv_f + 0.00523 \times a_pA \\ -0.26871\, a_pv_f + 0.08662 \times v_fA - 0.00315 \times n^2 + 0.06724 \times a_p^2 - 1.68634 \times A^2 \\ -70.7778 \times v_f^2 \end{aligned} \tag{15}$$

$$\begin{aligned} \text{MRR} = -16.03421 + 7.38367 \times a_p + 3.00296 \times v_f + 1.97321 \times A \\ -0.73966 \times n + 1.88326 \times na_p + 0.08325 \times nA - 0.05507 \times nv_f \\ +0.30681 \times a_pA - 0.110766\, a_pv_f + 0.50526 \times v_fA - 0.77360 \times n^2 \\ +0.53086 \times a_{p0}^2 - 0.66773 \times A^2 - 0.35677 \times v_f^2 \end{aligned} \tag{16}$$

According to Formulas (15) and (16), the combination of the interaction between design variables that has the greatest influence on the response factor Ra is na_p, followed by nA. For the interaction between design variables, the combination with the greatest impact on the response factor MRR is na_p, followed by v_fA. The response surface of the interaction between the design variables to the response factors is shown in Figure 13.

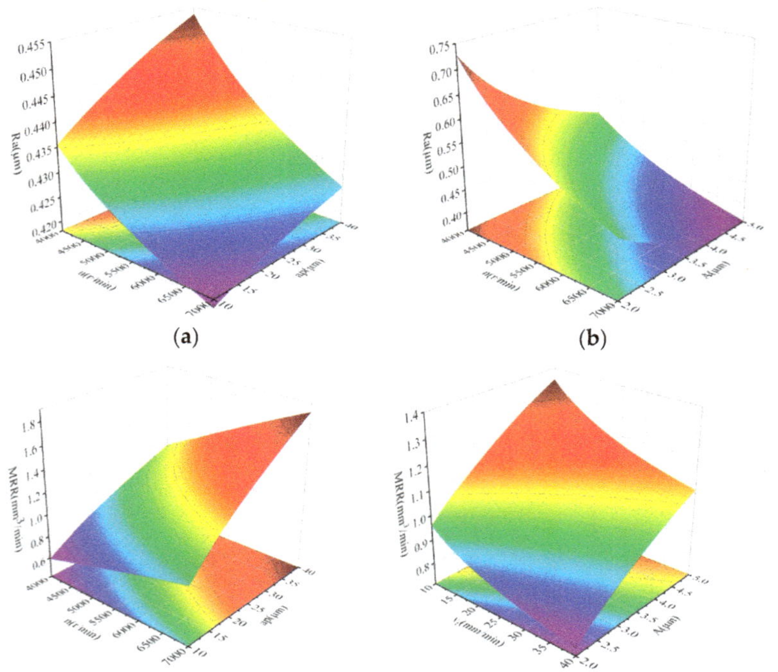

Figure 13. Response surfaces of Ra and MRR corresponding to process parameters. (**a**) Ra for n, a_p. (v_f = 30 mm/min, A = 3 μm). (**b**) Ra for n, A. (v_f = 30 mm/min, a_p = 40 μm). (**c**) MRR for n, a_p. (v_f = 30 mm/min, A = 3 μm). (**d**) MRR for v_f, A. (n = 5000 r/min, a_p = 30 μm).

Figure 13a is the response surface obtained under the condition of v_f = 30 mm/min and A = 3 μm. From the diagram, it can be seen that Ra increases with the increase in the grinding wheel speed and decreases with the increase in the grinding depth. Compared with the grinding depth, the influence of the grinding wheel speed on Ra is greater. Figure 13b is the response surface obtained under the condition of v_f = 30 mm/min and a_p = 40 μm. From the graph, it can be seen that Ra increases with the increase in the grinding wheel speed, and with the increase in the amplitude, the downward trend of Ra slows down. Compared with the amplitude, the influence of the grinding wheel speed on Ra is greater. Figure 13c is the response surface obtained under the condition of v_f = 30 mm/min and A = 3 μm. It can be seen from the figure that MRR increases with the increase in the grinding wheel speed and grinding depth. Compared with the grinding wheel speed, the grinding depth has a greater influence on Ra. Figure 13d is the response surface obtained under the condition of n = 5000 r/min and a_p = 30 μm. From the graph, it is known that MRR increases with the increase in amplitude A and decreases with the increase in the feed speed. From the contour line, the influence of the two parameters on MRR is small.

Aiming at low Ra and high MRR, the optimal combination of process parameters obtained by Design-Expert is:

A = 4.5 μm, n = 6493.3 r/min, a_p = 28.4 μm, v_f = 21.1 mm/min, predicted Ra = 0.472 μm, and MRR = 1.676 mm^3/min.

According to the optimal process parameters obtained by Design-Expert, the surface topography prediction model is introduced to simulate the surface topography, as shown in Figure 14. From the simulation diagram, it can be seen that the peak and trough are evenly distributed, and the peak and trough difference is small. The surface morphology and roughness of the longitudinal–torsional ultrasonic internal grinding test using the optimized parameters are shown in Figure 15. From the figure, it can be seen that the grinding surface morphology has good uniformity and is in good agreement with the simulated morphology. The measured surface roughness of the sample is Ra = 0.502 μm. According to the calculation, the relative error between the actual value of Ra and the predicted value is 5.9%. Because the roughness test and the material removal rate test use the same set of parameters, there is no need to carry out the second test. After calculation, the actual material removal rate is MRR = 1.763 mm^3/min, and the relative error between the actual value of MRR and the predicted value is 5.1%. The relative error between the actual value and the predicted value of the response factor shows that the optimization method is reliable and has guiding significance for the longitudinal–torsional ultrasonic internal grinding process of GCr15 bearing steel.

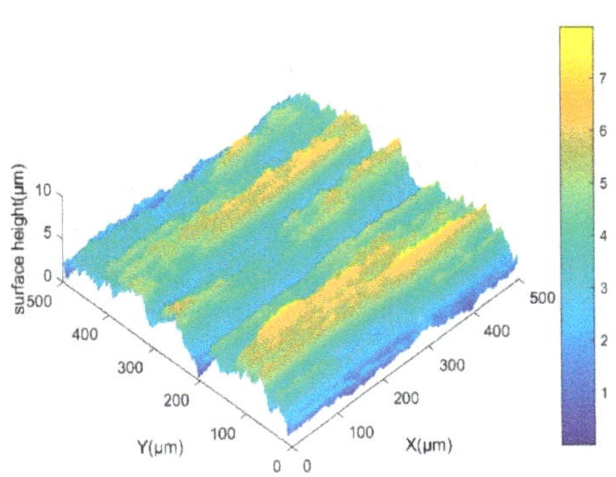

Figure 14. Optimized surface topography simulation.

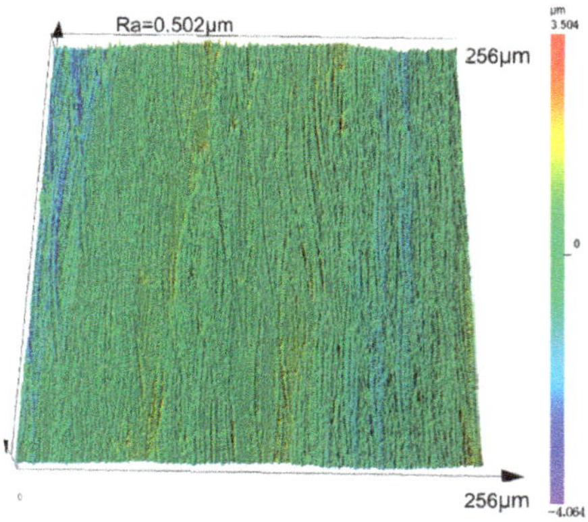

Figure 15. Surface morphology and roughness of the optimized samples.

5. Conclusions

In the presented work, a longitudinal–torsional ultrasonic grinding system was developed, the surface morphology prediction model of LTUG was established, and the process parameters were optimized. The main conclusions are as follows:

(1) A LTUG system was developed. The LUTG system ultrasonic frequency was 28.1 kHz, the longitudinal amplitude was between 0–5 μm, and the torsional amplitude had a longitudinal amplitude of 25%.

(2) Considering the established elastic–plastic deformation longitudinal–torsional ultrasonic grinding surface morphology prediction model of LUTG, the prediction accuracy was within 13.2% when compared with the test.

(3) Aiming at low Ra and high MRR Ra and MMR as the targets, the optimal process parameters were: A = 4.5 μm, n = 6493.3 r/min, a_p = 28.4 μm, and v_f = 21.1 mm/min.

Author Contributions: H.Z.: conceptualization, methodology, and writing—original draft; Y.N.: conceptualization, methodology, software, and supervision; X.J.: resources, supervision, and writing—review and editing; S.C.: software and investigation; J.N.: collected information. All authors have read and agreed to the published version of the manuscript.

Funding: This research was funded by the Henan Province's Key Research and Promotion (Scientific and Technological) Project (grant number 212102210335), the Science and Technology Project of Anyang City (grant number 2021C01SF056), and the Fundamental Research Funds for the Universities of Henan Province (NSFRF230409). Anyang Institute of Technology PhD initiation fund (BSJ2020007).

Data Availability Statement: The data are unavailable due to privacy.

Conflicts of Interest: No conflict of interest exits in the submission of this manuscript, and the manuscript has been approved by all authors for publication. I would like to declare on behalf of my coauthors that the work described was original research that has not been published previously and is not under consideration for publication elsewhere.

References

1. Giemza, B.; Nowiński, E.; Domański, M. Gas-Dynamic Foil Bearings Application in High-Speed Turbines. *J. Konbin* **2015**, *27*, 99–108. [CrossRef]
2. Xue, P.; Jiang, Y.; Wang, H.; He, H. Accurate Detection Method of Aviation Bearing Based on Local Characteristics. *Symmetry* **2019**, *11*, 1069. [CrossRef]
3. Schmidt, A.A.; Plánka, J.; Schmidt, T.; Grabherr, O.; Bartel, D. Validation of a dry sliding wear simulation method for wastegate bearings in automotive turbochargers. *Tribol. Int.* **2021**, *155*, 106711. [CrossRef]
4. Gao, X.; Yan, C.; Liu, Y.; Yan, P.; Yang, J.; Wu, L. A 4-DOF dynamic model for ball bearing with multiple defects on raceways. *Proc. Inst. Mech. Eng. Part K J. Multi-Body Dyn.* **2021**, *235*, 3–18. [CrossRef]
5. Liu, J.; Wu, H.; Shao, Y. The influence of the raceway thickness on the dynamic performances of a roller bearing. *J. Strain Anal. Eng. Des.* **2017**, *52*, 528–536. [CrossRef]

6. Li, Q.; Pan, C.; Jiao, Y.; Hu, K. Research on PCBN Tool Dry Cutting GCr15. *Machines* **2018**, *6*, 28. [CrossRef]
7. Manieri, F.; Stadler, K.; Morales-Espejel, G.E.; Kadiric, A. The origins of white etching cracks and their significance to rolling bearing failures. *Int. J. Fatigue* **2019**, *120*, 107–133. [CrossRef]
8. Arakere, N.K. Gigacycle rolling contact fatigue of bearing steels: A review. *Int. J. Fatigue* **2016**, *93*, 238–249. [CrossRef]
9. Kan, Z.; Wenhe, L.; Lianjun, S.; Heng, M. Investigation on grinding temperature in ultrasonic vibration-assisted grinding of zirconia ceramics. *Mach. Sci. Technol.* **2019**, *23*, 612–628. [CrossRef]
10. Xu, L.-H.; Na, H.-B.; Han, G.-C. Machinablity improvement with ultrasonic vibration–assisted micro-milling. *Adv. Mech. Eng.* **2018**, *10*, 168781401881253. [CrossRef]
11. Guo, M.; Lu, M.; Lin, J.; Gao, Q.; Du, Y. Modeling and investigation of minimum chip thickness for silicon carbide during quasi-intermittent vibration–assisted swing cutting. *Int. J. Adv. Manuf. Technol.* **2023**, *127*, 1691–1701. [CrossRef]
12. Wang, Z.; Niu, Y.; Sun, H.; Jiao, F. Preparation and mechanism of copper ultra-fine grain strip by ultrasonic vibration extrusion cutting-model, FEM and experiment. *J. Mater. Res. Technol.* **2023**, *24*, 4919–4934. [CrossRef]
13. Niu, Y.; Jiao, F.; Zhao, B.; Gao, G. Investigation of Cutting Force in Longitudinal-Torsional Ultrasonic-Assisted Milling of Ti-6Al-4V. *Materials* **2019**, *12*, 1955. [CrossRef]
14. Chen, Y.; Su, H.; He, J.; Qian, N.; Gu, J.; Xu, J.; Ding, K. The Effect of Torsional Vibration in Longitudinal–torsional Coupled Ultrasonic Vibration-Assisted Grinding of Silicon Carbide Ceramics. *Materials* **2021**, *14*, 688. [CrossRef] [PubMed]
15. Yang, Z.; Zhu, L.; Ni, C.; Ning, J. Investigation of surface topography formation mechanism based on abrasive-workpiece contact rate model in tangential ultrasonic vibration-assisted CBN grinding of ZrO_2 ceramics. *Int. J. Mech. Sci.* **2019**, *155*, 66–82. [CrossRef]
16. Gao, G.; Zhao, B.; Xiang, D.; Kong, Q. Research on the surface characteristics in ultrasonic grinding nano-zirconia ceramics. *J. Mater. Process. Technol.* **2009**, *209*, 32–37. [CrossRef]
17. Xiao, C.-F.; Han, B. Research and Design of Ultra-long Ultrasonic Horn. *J. Inst. Eng. Ser. C* **2019**, *100*, 1015–1022. [CrossRef]
18. Xiping, H.; Haidao, Z. Analytical and experimental study of an ultrasonic horn with a bend angle. *J. Vib. Control* **2018**, *24*, 4383–4394. [CrossRef]
19. Yang, J.; Ji, S.; Zhao, J.; He, Q. Theoretical analysis and finite element calculation of ultrasonic horn. *Mater. Sci. Eng.* **2019**, *612*, 032032. [CrossRef]
20. Hahn, M.; Cho, Y.; Jang, G.; Kim, B. Optimal Design and Experimental Verification of Ultrasonic Cutting Horn for Ceramic Composite Material. *Appl. Sci.* **2021**, *11*, 1954. [CrossRef]
21. Zhou, M.; Zheng, W. A model for grinding forces prediction in ultrasonic vibration assisted grinding of SiCp/Al composites. *Int. J. Adv. Manuf. Technol.* **2016**, *87*, 3211–3224. [CrossRef]
22. Hadhoum, L.; Loubar, K.; Paraschiv, M.; Burnens, G.; Awad, S.; Tazerout, M. Optimization of oleaginous seeds liquefaction using response surface methodology. *Biomass-Convers Biorefin.* **2020**, *11*, 2655–2667. [CrossRef]
23. Hattab, M.W. On the use of data transformation in response surface methodology: A note on data transformation in response surface methodology. *Qual. Reliab. Eng. Int.* **2018**, *34*, 1185–1194. [CrossRef]
24. Zhang, Y.; Li, C.; Chu, D.; Yan, G.; Zhu, M.; Zhao, X.; Gu, J.; Li, G.; Wang, J.; Zhang, B. Process optimization for the preparation of thiamethoxam microspheres by response surface methodology. *React. Funct. Polym.* **2020**, *147*, 104460. [CrossRef]

Disclaimer/Publisher's Note: The statements, opinions and data contained in all publications are solely those of the individual author(s) and contributor(s) and not of MDPI and/or the editor(s). MDPI and/or the editor(s) disclaim responsibility for any injury to people or property resulting from any ideas, methods, instructions or products referred to in the content.

micromachines

MDPI

Article

Process Analysis and Topography Evaluation for Monocrystalline Silicon Laser Cutting-Off

Fei Liu [1], Aiwu Yu [2], Chongjun Wu [1,*] and Steven Y. Liang [3]

1 College of Mechanical Engineering, Donghua University, Shanghai 201620, China; lf2871157603@163.com
2 Shanghai Aerospace Equipment Manufacturer Co., Ltd., Shanghai 201109, China; calebyu1988@163.com
3 Manufacturing Research Center, Georgia Institute of Technology, Atlanta, GA 30332, USA; steven.liang@me.gatech.edu
* Correspondence: wcjunm@dhu.edu.cn

Abstract: Due to the characteristics of high brittleness and low fracture toughness of monocrystalline silicon, its high precision and high-quality cutting have great challenges. Aiming at the urgent need of wafer cutting with high efficiency, this paper investigates the influence law of different laser processes on the size of the groove and the machining affected zone of laser cutting. The experimental results show that when laser cutting monocrystalline silicon, in addition to generating a groove, there will also be a machining affected zone on both sides of the groove and the size of both will directly affect the cutting quality. After wiping the thermal products generated by cutting on the material surface, the machining affected zone and the recast layer in the cutting seam can basically be eliminated to generate a wider cutting seam and the surface after wiping is basically the same as that before cutting. Increasing the laser cutting times will increase the width of the material's machining affected zone and the groove width after chip removal. When the cutting times are less than 80, increasing the cutting times will increase the groove width at the same time; but, after the cutting times exceed 80, the groove width abruptly decreases and then slowly increases. In addition, the lower the laser scanning speed, the larger the width of the material's machining affected zone and the width of the groove after chip removal. The increase in laser frequency will increase the crack width and the crack width after chip removal but decrease the machining affected zone width. The laser pulse width has a certain effect on the cutting quality but it does not show regularity. When the pulse width is 0.3 ns the cutting quality is the best and when the pulse width is 0.15 ns the cutting quality is the worst.

Keywords: monocrystalline silicon; laser cutting; ablation zone; quality analysis

1. Introduction

In recent years, semiconductor technologies have developed rapidly. Among them, monocrystalline silicon is a typical orthotropic hard brittle semiconductor material [1–3] which has excellent thermal conductivity, mechanical strength, overload resistance, and other properties [4,5]. It has been widely used in aerospace, national defense construction, biotechnology, and other aspects [6–8]. However, the high brittleness and low fracture toughness of monocrystalline silicon make it a typically difficult to machine material [5,9,10].

The initial method for cutting crystalline silicon is mortar wire cutting but this method has a low cutting efficiency, high material loss, and a certain degree of water pollution so it is gradually replaced by diamond wire cutting. The diamond wire-cutting method has a high cutting efficiency and low silicon material loss [11] so it has become one of the key technologies in the solar cell manufacturing process and semiconductor chip manufacturing process [12,13]. Costa et al. [14] discussed the influence of diamond wire cutting on the feed force and the quality of monocrystalline silicon wafers, found a method to reduce the feed force, and obtained smoother material surface and shallower microcracks. Costa et al. [15] investigated the effect of diamond wire saws on the surface integrity of monocrystalline silicon and used circular diamond wire saws to cut the monocrystalline silicon. They found that the most suitable cutting parameters were the lowest feed rate and wire tension and the highest wire cutting speed. Wang et al. [16] investigated the effect of the scribing speed on the surface morphology and material removal behavior of monocrystalline silicon diamond wire saws by designing high-speed diamond scribing experiments and explained the potential mechanism of promoting brittle fracture at higher scratching speeds. But the cutting efficiency of traditional diamond wire cutting is still not high; it is far from meeting the market demand.

With the development of science and technology, laser cutting technology is becoming more and more mature [17]. Laser cutting technology is one of the most widely used non-contact material cutting methods [18,19] which has the advantages of a narrow groove, small machining affected zone, fast cutting speed, higher precision, and good controllability [20,21]. Zhao et al. [22] used a laser to cut anisotropic monocrystalline silicon bilayer wafers and found that the anisotropy of monocrystalline silicon had a great impact on the quality of crack edge morphology and the form of silicon fracture after cutting. Mulugeta et al. [23] used laser technology to cut silicon anodes and determined the minimum average power and energy efficiency of four cutting widths and five cutting methods.

However, due to the immature laser cutting technology, the processing effect of monocrystalline silicon is not ideal [24,25], especially since the machining affected zone cannot be completely eliminated in the laser direct cutting [26] and the excessive machining affected zone will reduce the chip performance. Huang et al. [27] used picosecond lasers with different fluences to cut ultra-thin wafers. At high fluence, the heat-affected zone only exists at the edge position. At low flux, more crystallographic defects were found at the edge of the heat-affected zone. In contrast, the laser-induced thermal crack propagation is considered to be a promising process for silicon cutting [28–30]. This is because it has the advantage of avoiding these problems [31–33] and producing a good surface finish [34,35].

However, there is limited research on the impact of laser processing parameters on cutting quality and optimizing relevant parameters can effectively improve the wafer cutting quality. Therefore, this paper is aimed at the low ablation area of laser wafer cutting and the urgent need for high-efficiency wafer cutting to carry out the research of laser wafer cutting technology. A laser cutting experiment of single crystal silicon was designed with the basic requirement of reducing the laser cutting times and the main objective of optimizing the parameters of laser scanning speed, frequency, and pulse width.

2. Experiment Setup

2.1. Materials and Methods

2.1.1. Wafer Material

The material used in this experiment was single-sided polished silicon wafer which is a kind of brittle and hard material. After the silicon crystal was crystallized and grown by the straight pulling method, one of the surfaces was in a mirror and shiny state after mechanical or chemical treatment. Monocrystalline silicon is a kind of crystal with a basically complete lattice structure. It has different properties in different crystal directions and is a good semi-conductive material. The surface of a complete silicon chip is like a mirror and the overall shape is like a circle with a small part cut off on the perimeter, hence it is also called a wafer. The direction of the tangent line is the direction of the 0° crystal direction. Table 1 shows the main parameters of the chip used in the experiment.

Table 1. Wafer Parameters.

Basic Parameter			
Type	Single side polished silicon wafer	**Thickness**	200 ± 10 μm
Model/crystal orientation	P<100>	**Diameter**	100 ± 0.4 mm
Growth method	CZ	**Resistivity**	1–10 Ωcm

2.1.2. Wafer-Cutting Morphology and the Judgment Method

This experiment aimed to explore the influence of laser cutting on the surface quality and thermal damage of wafer materials in order to obtain higher quality and less damaged wafers. Similarly to traditional cutting, laser cutting makes use of the high-temperature characteristics of laser cutting to rapidly melt the material at the cutting position and sputter outward, leaving a fine line on the surface of the material. With the increase in cutting times, the fine line gradually moved to the inside of the material, just like the feeding in the turning process, until the bottom of the material is cut off. At the end of each cut, the focus of the laser beam was repositioned to the bottom of the material to begin the next cut. Figure 1 gives the measured view for the wafer cutting.

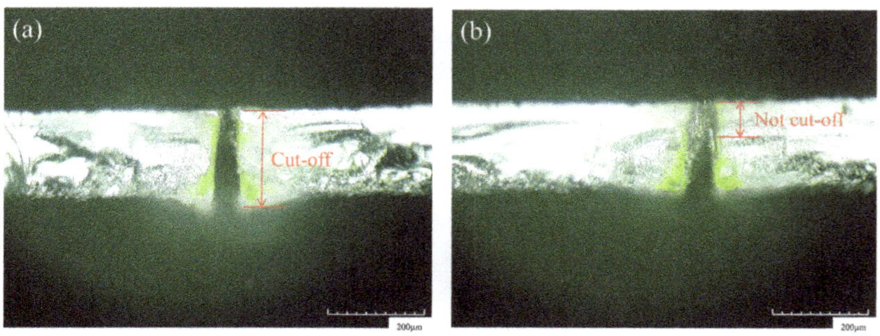

Figure 1. Schematic diagram of whether the material is cut-off or not. (**a**) Cut-off and (**b**) not cut-off.

2.1.3. Laser Processing Equipment

This experiment used a Huaray Polar-355 nanosecond laser for the wafer-cutting experiment, the detailed parameters are given in Table 2. The laser is a pulse laser which can increase and emit laser energy in a very short time as well as through a high frequency of light to complete the material cutting. Due to the light weight of the silicon wafer and the thermal deformation phenomenon during the cutting process, the cutting positions of the silicon wafer vary with different cutting times. Therefore, during the experiment, it was necessary to fix the material to ensure that each cut was carried out on one groove.

Table 2. Parameters of the Huaray Polar-355 nanosecond laser.

Basic Parameters			
Output power	>5 W@50 kHz	Working temperature	10–35 °C
Single pulse energy	>125 μJ@40 kHz	Cooling method	Water
Pulse width	16 ± 2 ns@50 kHz	Voltage	110/220 V, 50/60 Hz
Start time	<15 min	Focus (z-axis direction)	33 cm
Repetition frequency	20 kHz–200 kHz	Spatial mode	TEM_{00} ($M^2 \leq 1.2$)
Power stability	≤3% rms	Pulse stability	≤3% rms
Beam diameter	<8 m	Spot divergence angle	≤1 mrad
Beam directivity	<25 μrad	Rad spot roundness	≥90%
Manufacturer	Wuhan Huaray Precision Laser Co., Ltd.	Nation	Wuhan, China

2.1.4. Selection of the Measuring Instrument

In this experiment, the Hirox KH-7700 digital 3D video microscope was selected as the parameter-measuring instrument. The instrument uses optical zoom to observe the 2D image of the selected altitude layer and can obtain the 3D vertical image of the selected point through multi-layer superposition technology and Z-axis autonomous movement so as to measure the relevant parameters required for the experiment. The measuring instrument is shown in Figure 2.

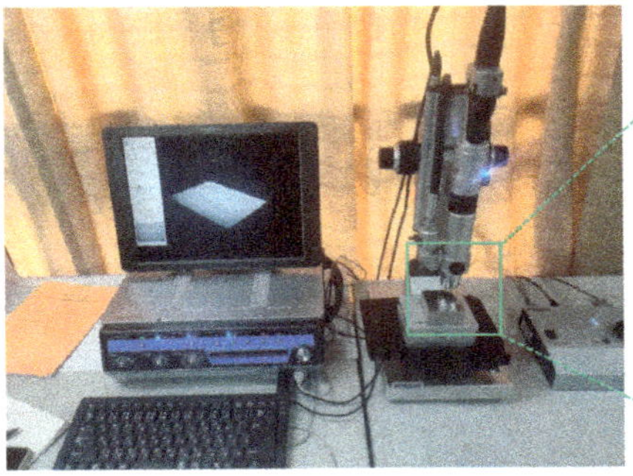

(a) 3D microscope

(b) Enlarged zone

Figure 2. Measuring instrument.

2.2. Laser Cutting Process Parameters Selection

The materials used in this study included single-sided polished silicon wafers which are prone to heat damage in the process of laser cutting. Therefore, this experiment mainly studied the influence of different laser process parameters on the surface quality and thermal damage size of the wafer after laser cutting and finally completed the optimization of laser parameters. Considering the requirements of mass production for material quality, the production efficiency in the final application, and the linear correlation between production time and cutting times, with the increase in cutting times the surface machining affected zone also increased, so the cutting times were taken as the primary research object to find the minimum cutting times that can cut the material exactly.

In this experiment, the first experiment was the cutting times as the independent variable and other parameters such as laser frequency and scanning speed remained unchanged. Laser parameters: scanning speed: 200 mm/s, laser frequency: 40 kHz, pulse width: 0.3 ns, cutting times: 5, 10, 15, 20, 30, . . ., 200, 300 times. The minimum cutting times were obtained when the wafer was cut off.

Due to the previously set experimental parameters ranging from 5 to 300 cuts, with a large parameter span, in order to ensure the accuracy of the experimental results, after determining the optimal cutting times (X) that meet the cutting requirements, based on X, 11 sets of laser cutting experiments were added to further determine the minimum cutting times.

After finding the minimum cutting times that meet the experimental requirements, a control experiment was designed around the cutting times to explore the impact of different laser processes on the damage of cutting quality. The specific experimental parameter design is shown in Table 3.

Table 3. Experimental test and analysis.

	Parameters
Scanning speed/mm·s^{-1}	50, 100, 150, 200, 250, 300, 350
Laser frequency/kHz	20, 30, 40, 50, 60
Pulse width/ns	0.15, 0.2, 0.25, 0.3, 0.35, 0.4, 0.45
Laser passes	20

2.3. Overall Experiments Design

The process of laser cutting wafer is essentially a process of drilling which belongs to the vaporization cutting. Lasers have high energy and power. Taking advantage of these characteristics, laser energy accumulates massively on the wafer surface, the material rapidly heats up and vaporizes, some materials are blown away as slag by auxiliary gas, the relative movement of laser and wafer forms a groove, and the laser generates energy accumulation and heat conduction on the wafer surface, leading to the formation of machining affected zones on both sides of the groove. Figure 3 illustrates the development of Wafer Cutting.

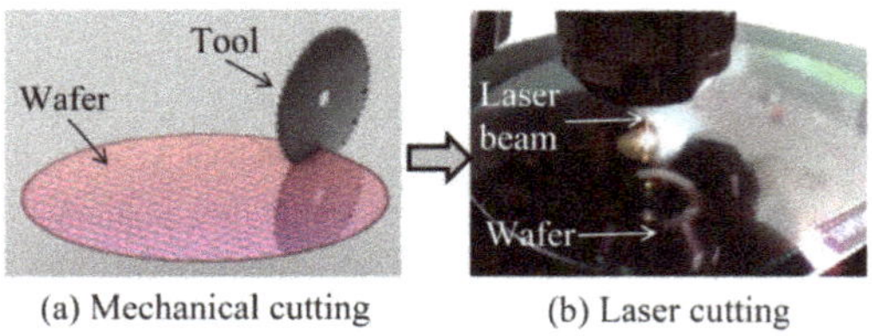

Figure 3. Development of Wafer Cutting.

As the surface parameters can be obtained directly from laser cutting experiments, the width of the groove and the width of the machining affected zone directly reflect the quality of laser cutting. By sorting out and analyzing the material surface parameters obtained from the previous laser cutting experiment, we can obtain the influence of the cutting times and laser process parameters on the wafer cutting.

During the experiment, by wiping the material surface after cutting, it was found that the machining affected zone on the material surface and the recast layer near the surface on both sides of the groove could be directly removed. In the actual production, if you can use this phenomenon you can increase the utilization rate of materials and reduce production costs, hence the width of the cutting seam and the width of the morphology after the removal of chips are the research objects in order to explore the influence of different laser parameters on its rule. The experimental process is shown in Figures 4 and 5.

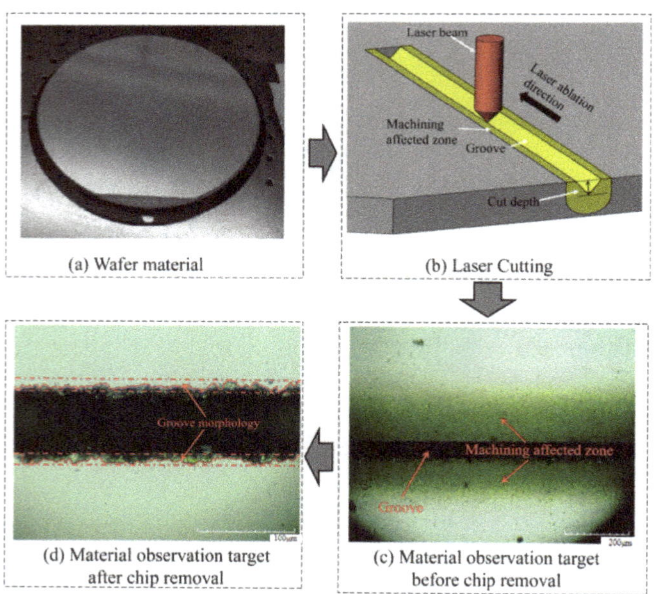

Figure 4. Laser cutting and the observation target.

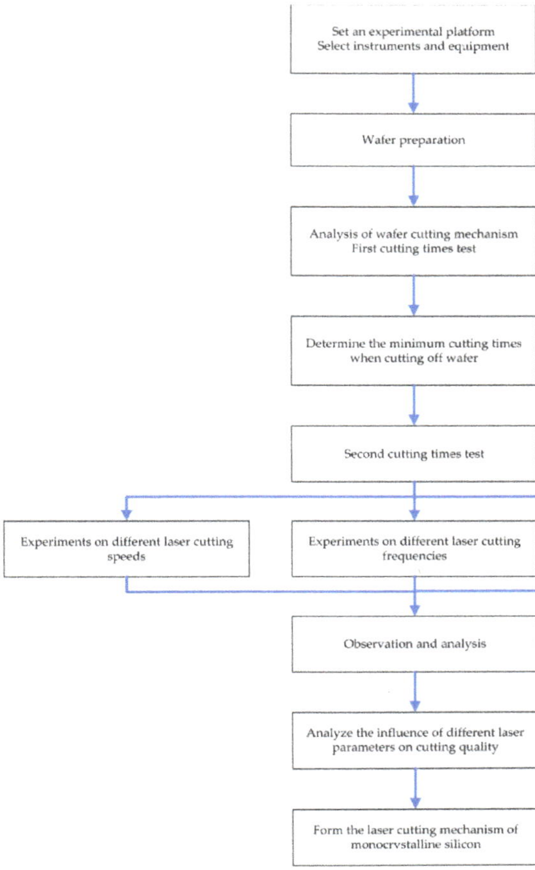

Figure 5. Schematic Diagram.

This paper focused on the quality control and damage analysis of laser cutting of monocrystalline silicon. In this experiment, firstly, an experimental platform should be established and the appropriate instruments selected. Then, the monocrystalline silicon wafer should be pretreated. Through the laser cutting wafer experiment, the materials should be cut with different cutting times and laser parameters in order to analyze the morphology of the material surface and the machining affected zone. Finally, the laser-cutting mechanism of monocrystalline silicon should be formed.

3. Analysis of the Wafer Cutting Mechanism

3.1. Wafer Cutting Process Analysis

3.1.1. Judgment of the Threshold Value of Laser Cutting Times

According to the previous experimental design, the initial experiment was carried out with the cutting times as the independent variable and the laser frequency, speed, and other parameters remained unchanged. The parameter values not given in the table are the scanning speed: 200 mm/s; laser frequency: 40 kHz; and pulse width: 0.3 ns. After the laser cutting was completed, the material surface was observed with a three-dimensional microscope and then an alcoholic cotton ball was used to wipe both sides of the groove to remove the chips; after observation of the chips this process was repeated if necessary. The data obtained were sorted out.

The experiment results show that the materials were cut off with the increase in cutting times from the 20th time. Hence, this experiment shows that the cutting times of about 20 times are the lowest times of material cutting. The experiment contents of the laser cutting experiment are shown in Table 4 below.

Table 4. First experiment parameters are based on the change in cutting times.

Laser Passes	Crack Width/μm	Width of Machining Affected Zone/μm	Crack Width after Chip Removal/μm	Whether It Is Cut Off
5	53	286	155	No
10	57	330	188	No
15	62	356	206	No
20	64	355	211	Yes
30	66	349	223	Yes
40	67	349	247	Yes
60	65	380	250	Yes
80	54	356	242	Yes
120	50	355	236	Yes
160	50	356	272	Yes
200	49	349	253	Yes
300	52	359	272	Yes

3.1.2. Determination of Laser Cutting Times

Based on the results of the first experiment for 20 times, we added 11 groups of laser cutting experiments and identified the burn conditions. The identification of the thermal damage and cutting conditions of six groups of laser cutting experiments in which the cutting times were even shows that with the increase in the test times, the material's groove width, the width of the machining affected zone, and the groove width after chip removal have an increasing trend. After observing the crack after chip removal, it was found that both sides of the original groove with the recast layer were more uniform and closer to a straight line, while both sides of the groove after chip removal were uneven and closer to a broken line. Moreover, the range of broken lines corresponding to different laser parameters was different, so the groove morphology after chip removal was taken as another research object for experimental reference. The specific experimental parameters and cutoff conditions are shown in Table 5.

Table 5. Parameters of the second experiment based on the change in cutting times.

Laser Passes	Crack Width/μm	Width of Machining Affected Zone/μm	Crack Width after Chip Removal/μm	Width of Groove Morphology/μm	Whether It Is Cut Off
16	62	348	69	24	No
18	64	354	72	24	No
20	64	355	72	23	Yes
22	67	363	74	27	Yes
24	65	354	72	28	Yes
26	70	356	76	20	Yes

3.1.3. Analysis of Laser Cutting Quality

Analysis of the above data shows that with the increase in cutting times, the cutting depth also increases, and when the cutting times is 20 times, the material is completely cut off. This is because cutting laser energy is limited: a single cut cannot directly penetrate through the material. If lasers irradiate the surface of the material every time, it will feed a distance into the material until cut off. The following figures show the material morphology under different cutting times. Figure 6 shows the morphology of the material before chip removal under different cutting times.

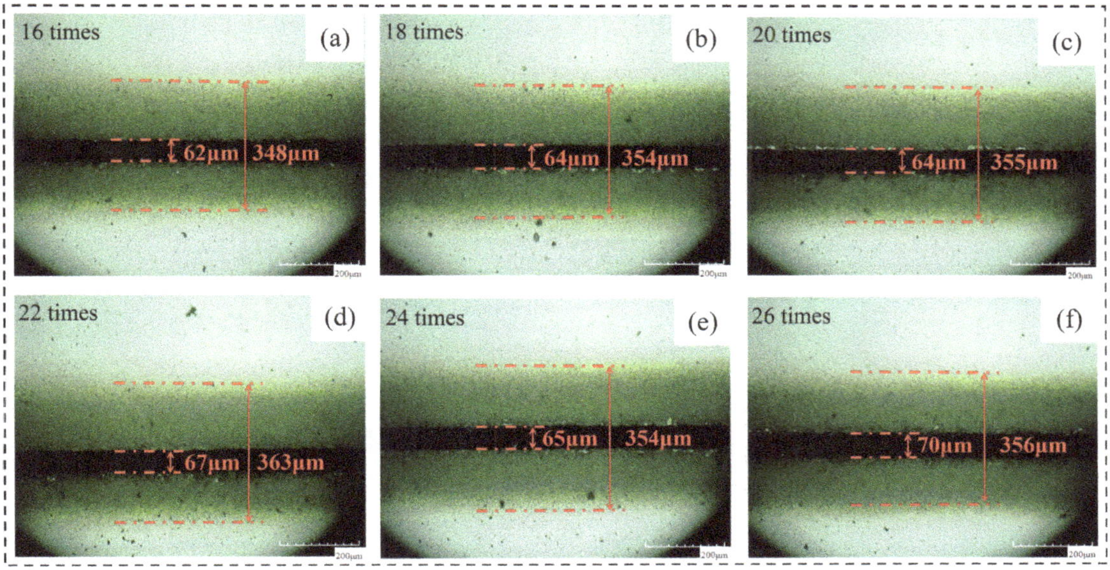

Figure 6. The material morphology before chip removal. (**a**) 16 times, (**b**) 18 times, (**c**) 20 times, (**d**) 22 times, (**e**) 24 times, (**f**) 26 times.

3.2. Micromorphology and Machining Affected Zone Analysis

3.2.1. Material Crack Width Analysis

With the change in laser cutting times, the crack width on the material surface will change. Figure 7 shows the changing trend of the crack width on the material surface with the change in cutting times. Through analysis, the influence rule of laser cutting times on the crack width was obtained: when the laser cutting times are low (less than 80 times), the width of the groove increases with the increase in the cutting times, but when the laser cutting times exceed 80 times, the groove width will rapidly decrease and again slowly increase. This is because the process of laser cutting is a process of melting, decomposition, and the vaporization of materials as well as the spraying and solidification of molten materials. Finally, a certain thickness of the recast layer will be formed on both sides of the cutting seam. When the cutting times are low, the width of the cutting seam is less affected by the thickness of the recast layer so it gradually increases with the cutting times. When the cutting times reach a certain value, at this time, the groove width of laser cutting is more affected by the width of the recast layer than by the direct impact of laser cutting so the groove width decreases with the increase in the laser cutting times. If the cutting times continue to increase, the temperature rises, the recast layer material vaporizes, the thickness becomes smaller, and then the groove width increases again. Figure 8 shows the change in surface crack width of the material as the cutting times increase.

3.2.2. Material Machining Affected Zone Analysis

With the change in laser cutting times, the width of the machining affected zone will also change. Figure 9 shows the trend of the width of the machining affected zone on the material surface changing with the cutting times. Through analysis, the influence rule of laser cutting times on the crack width is obtained: with the increase in laser cutting times, the width of the machining affected zone of materials gradually increases and finally tends to be stable. This is because when the laser cutting times are low, the surface temperature of the material increases with the increase in the cutting times, and gradually extends along the material surface to both sides of the cutting seam. When the cutting times reach a certain value, even if the temperature at the material cutting position continues to increase with the cutting times, the machining affected zone will not continue to extend to both sides due to the thermal conductivity of the material itself, so the width of the machining affected zone at the edge position tends to be stable.

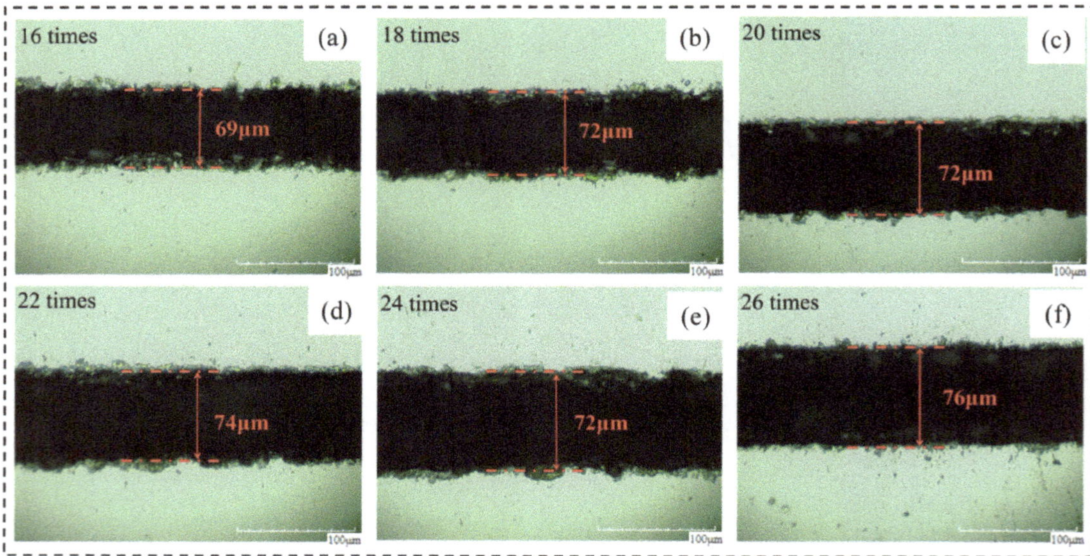

Figure 7. The material morphology after chip removal. (**a**) 16 times, (**b**) 18 times, (**c**) 20 times, (**d**) 22 times, (**e**) 24 times, (**f**) 26 times.

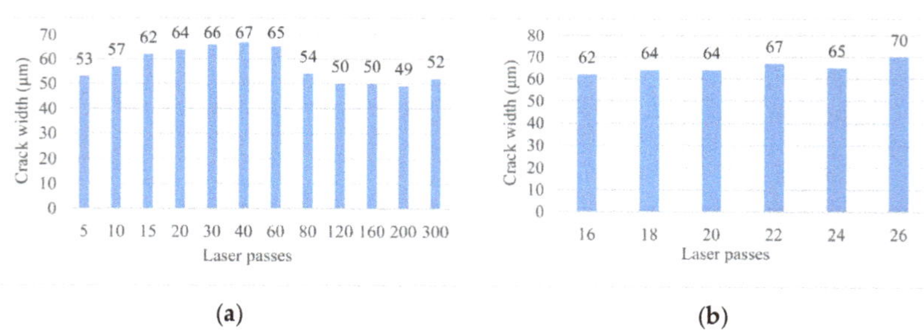

Figure 8. Curve of the surface crack width changing with cutting times. (**a**) Crack width after 5–300 times of cutting and (**b**) crack width after 16–26 times of cutting.

3.2.3. Analysis of Crack Width and Morphology after Chip Removal

Figure 10 shows the crack width of the material after chip removal on the material surface changing with the cutting times. By analyzing the surface morphology of the material after chip removal, the following conclusions were obtained. (1) Crack width after chip removal: the crack width after chip removal increases with the increase in cutting times, which is different from the rule that the crack width without chip removal is affected by the change in cutting times. This is because the use of alcohol cotton balls not only erased the machining affected zone on both sides of the groove but also erased the recast layer near the material surface at the material notch. Therefore, the groove width at this time is only affected by the laser energy received during the experiment and is positively related to it. (2) Width of groove morphology after chip removal: The change in cutting times will have a certain impact on the width of groove morphology after chip removal but the rule is not significant enough. Mainly because there are fewer cutting times and the impact of chips is relatively small. When the cutting times are 20, the crack width is small.

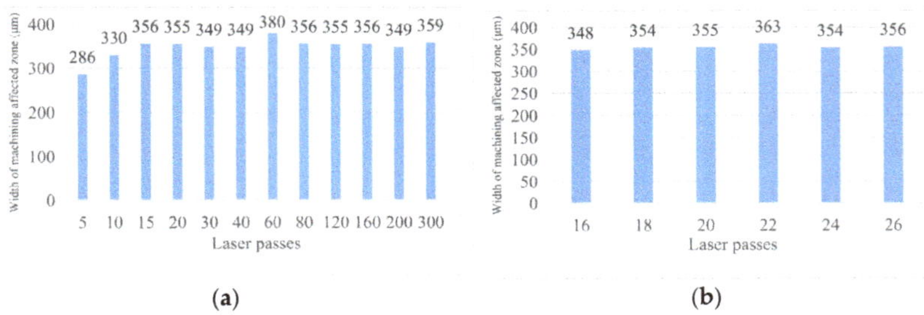

Figure 9. Curve of the width of the machining affected zone changes with cutting times. (**a**) Cutting 5–300 times and (**b**) cutting 16–26 times.

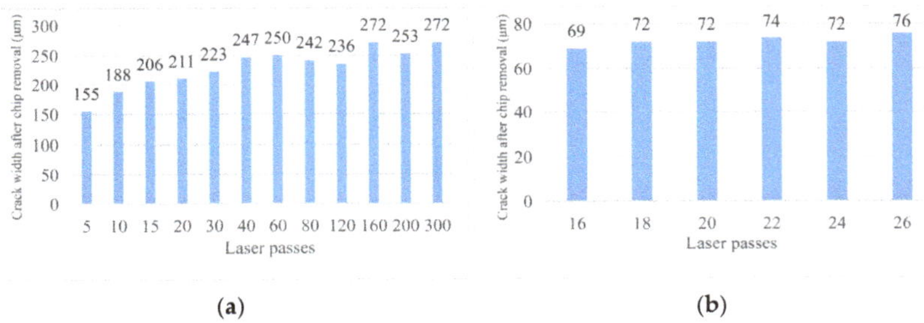

Figure 10. Crack width of the material after chip removal. (**a**) Cutting 5–300 times and (**b**) cutting 16–26 times.

4. Laser Experiment and Analysis

4.1. Process Analysis of the Influence of Laser Cutting Speed

4.1.1. Process Parameter Design

According to the previous experimental results, 20 times of laser cutting is the best experimental parameter. Keep other process parameters unchanged and the cutting times is 20. Carry out Scientific control with the laser speed as an independent variable. Use a three-dimensional microscope to analyze and observe the cut material surface, and process and summarize the data. The specific experimental results are shown in Table 6.

Table 6. Experimental data based on the scanning speed.

Scanning Speed/mm·s^{-1}	Crack Width/μm	Width of Machining Affected Zone/μm	Crack Width after Chip Removal/μm	Width of Groove Morphology/μm	Whether It Is Cut Off
50	60	351	79	23	Yes
100	70	371	78	24	Yes
150	69	354	75	23	Yes
200	64	355	72	23	Yes
250	60	354	67	21	Yes
300	58	342	64	32	No
350	60	331	65	29	No

4.1.2. Morphology Analysis before Chip Removal

By adjusting the laser scanning speed on the crack width and machining, the affected zone width was analyzed and the following conclusion was obtained. (1) Cutting effect:

the cutting times under the condition of fixed for 20 times, the lower the laser scanning speed is, the more obvious the cutting effect is. This is because the laser is not continuous but has pulse properties. Keeping the cutting times unchanged, the slower the scanning speed is, the more times the material is scanned per unit length, and the greater the cutting depth is. (2) Crack width: when the laser scanning speed exceeds 100 mm/s, the groove width gradually decreases with the increase in the laser scanning speed but when the laser scanning speed is too small (50 mm/s), the groove width also has a decreasing trend. This is because when the laser frequency is unchanged, the number of laser pulses emitted per unit time is the same, the laser scanning speed increases, the number of pulses per unit length of the material decreases, so the groove width becomes smaller. However, when the scanning speed is too small, the material temperature increases, the thickness of recast layer increases, and the groove width becomes smaller. (3) Width of machining affected zone: with the increase in scanning speed, the width of the material's machining affected zone gradually decreases. The reason is the same as the analysis of the effect of the laser scanning speed on the groove width. Figure 11 shows the changing trend of the relevant area with the scanning speed and Figure 12 shows the material morphology at a partial scanning speed.

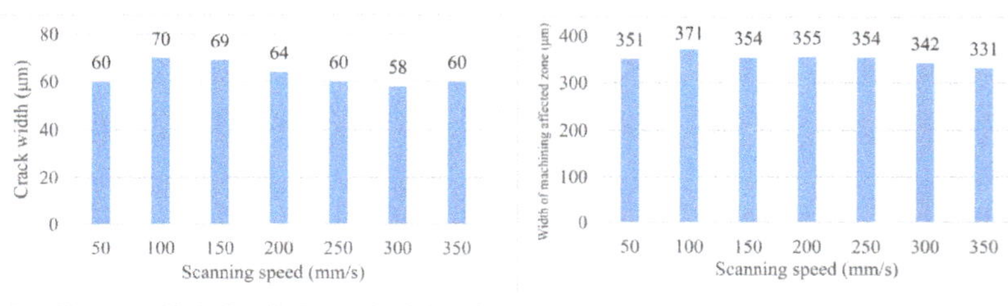

Figure 11. Change in material morphology before chip removal in speed experiment. (**a**) Material crack width. (**b**) material machining affected zone width.

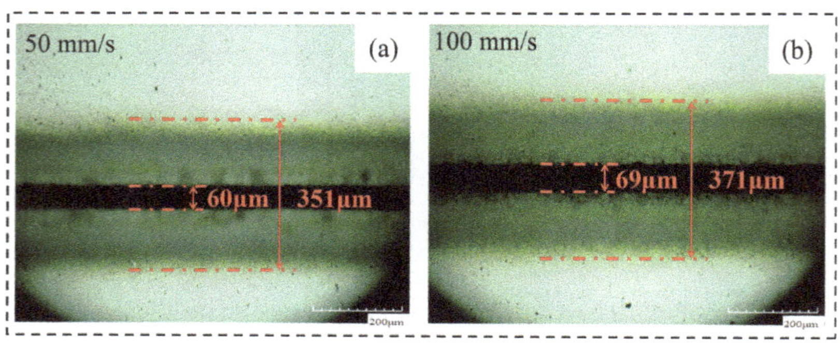

Figure 12. Material morphology at a partial cutting speed. (**a**) 50 mm/s, (**b**) 100 mm/s.

4.1.3. Morphology Analysis after Chip Removal

By analyzing the width of the chip morphology and groove morphology at different laser scanning speeds, the following conclusions are obtained. (1) Crack width after chip removal: the crack width after chip removal decreases with the increase in scanning speed because the alcohol cotton ball not only erases the machining affected zone on both sides of the groove but also erases the recast layer at the material notch; at this time, the groove width is only affected by the laser energy during the experiment and is positively correlated with it. (2) Groove profile width: the groove profile width increases with the increase in scanning speed and decreases with the decrease in scanning speed. Figure 13 shows the changing trend of the relevant area with the laser scanning speed. Figure 14 shows the (a) groove width and (b) groove morphology after cutting at a partial laser speed.

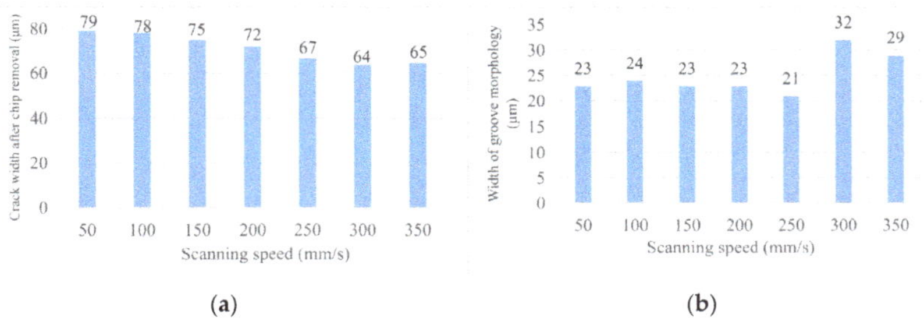

Figure 13. Change in material morphology after chip removal in speed experiment. (**a**) Groove width after chip removal and (**b**) groove morphology width.

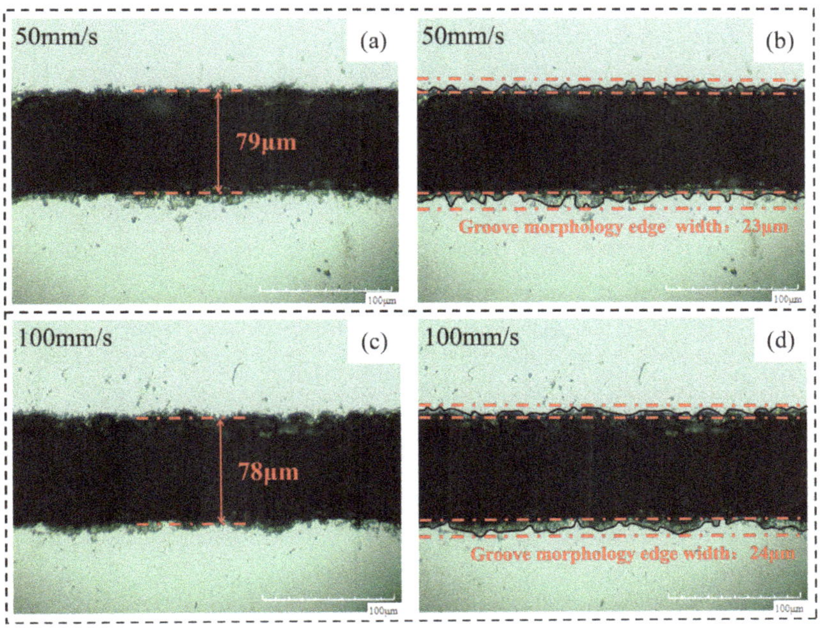

Figure 14. Material morphology after chip removal at partial cutting speed. (**a**) 50 mm/s groove width, (**b**) 50 mm/s groove morphology edge width, (**c**) 100 mm/s groove width, (**d**) 100 mm/s groove morphology edge width.

4.2. *Process Analysis of the Influence of Laser Frequency*

4.2.1. Process Parameter Design

According to the previous experimental results, the number of cutting times was determined to be 20, and other process parameters remained unchanged. The Scientific control was carried out with the laser frequency as the independent variable. In the process of the experiment, the ordinary laser mean power is equal to the product of the pulse energy and frequency. The formula for the laser used in this experiment can be used with the corresponding laser mean power frequency. Table 7 shows their corresponding relationships.

Table 7. Laser frequency and power.

Frequency/kHz	20	30	40	50	60
Average power/W	5	10	15	20	25

Therefore, when laser frequency is involved in the table, the unit name used is laser power/W (frequency/kHz). A three-dimensional microscope was used to observe the cut material surface and sort out the data obtained, as shown in Table 8.

Table 8. Experimental data based on laser frequency.

Power/W (Frequency/kHz)	Crack Width/μm	Width of Machining Affected Zone/μm	Crack Width after Chip Removal/μm	Width of Groove Morphology/μm	Cut Off
5 (20)	51	383	57	23	No
10 (30)	58	371	66	20	Yes
15 (40)	64	355	72	23	Yes
20 (50)	71	337	72	22	Yes
25 (60)	69	285	73	20	Yes

4.2.2. Morphology Analysis before Chip Removal

Through the analysis of the crack width and machining affected zone width under different laser frequencies, the following conclusions can be obtained. (1) Cutting effect: when the cutting times are 20 times, the higher the laser frequency, the more obvious the cutting effect. This is because when the cutting time is unchanged, the higher the laser frequency, the more times the material is pulsed by the laser per unit length, and the greater the cutting depth. (2) Crack width: with the increase in laser frequency, the groove width of the material gradually increases. This is because of the average power of the laser = laser frequency * single pulse energy. When the single pulse energy of the laser remains unchanged, the frequency is increased. If the energy received by the material in unit time increases, the groove width increases. (3) Width of the machining affected zone: with the increase in laser frequency, the width of the machining affected zone of the material shows a trend of decreasing gradually. The peak intensity at the center of the spot is inversely proportional to the laser pulse frequency. The calculation method of the laser pulse energy (E_p) is the following:

$$E_p = \frac{\pi}{8}\tau_p d_p^2 I_p \tag{1}$$

Therefore, as the laser frequency increases, the peak intensity (I_p) at the center of the spot decreases. At this time, the laser pulse energy (E_p) decreases, the heating ability of the laser weakens, and the liquid molten material ejected during the laser cutting process decreases. So even if the increase in laser frequency will increase the number of lasers of material per unit length, this effect does not surpass caused by laser heating ability reduce spewed molten material to reduce the effects of the liquid, so with the increase in laser frequency, but the machining affected zone width decreases. Figure 15 shows the changing trend of the relevant region with the laser frequency. Figure 16 Material morphology at partial laser frequencies (a) material morphology before chip removal (b) amplification at the groove.

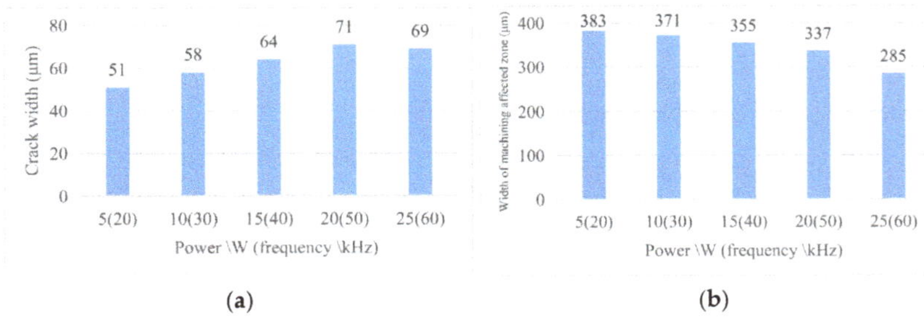

Figure 15. Change in material morphology before chip removal in frequency experiment. (**a**) Material crack width and (**b**) material machining affected zone width.

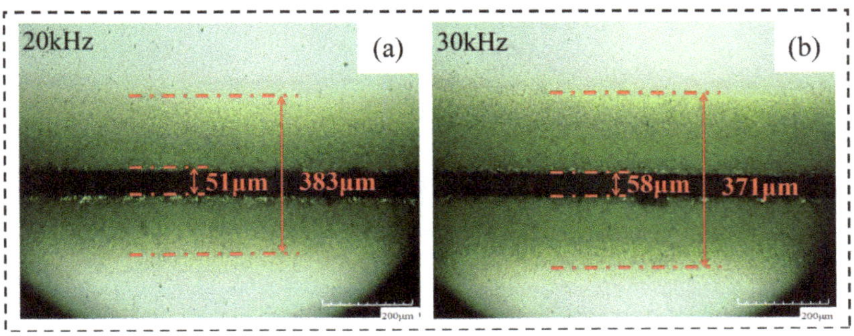

Figure 16. Material morphology at partial laser frequencies. (**a**) 20 kHz, (**b**) 30 kHz.

4.2.3. Morphology Analysis after Chip Removal

Through the analysis of the morphology width and groove morphology width after chip removal at different laser speeds, the following conclusions were obtained under the current experimental conditions. (1) Crack width after chip removal: For the effect of laser frequency on the crack width after chip removal, the crack width after chip removal increases with the increase in power; this is because with the laser parameters according to the trend change, the laser energy absorbed on the unit length of the groove increases, the local temperature increases, the material continues to melt, so the groove increases. (2) Groove morphology width: groove morphology width decreases with the increase in laser frequency and power and vice versa. Figure 17 shows the changing trend of the relevant region with the laser velocity and Figure 18 shows (a) the width of the groove and (b) the morphology of the groove after the chip removal at part of the laser frequency.

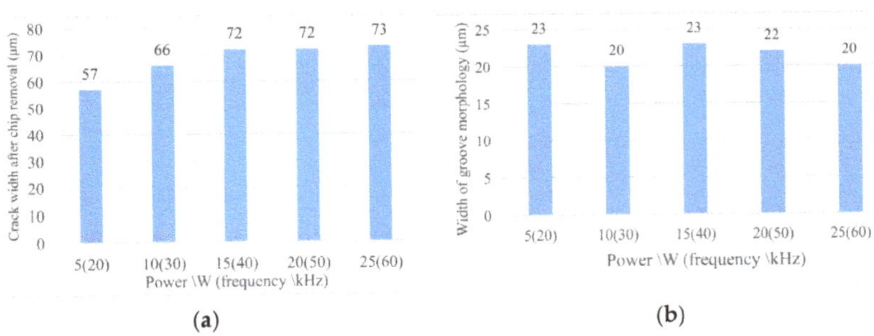

Figure 17. Change in material morphology after chip removal in frequency experiment. (**a**) Groove width after chip removal and (**b**) groove morphology width.

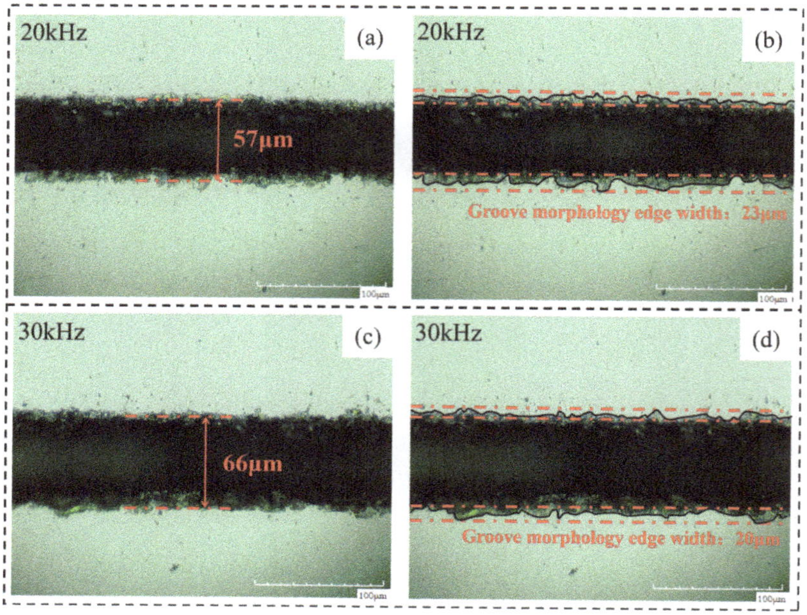

Figure 18. Material morphology after chip removal at partial laser frequencies. (**a**) 20 kHz groove width, (**b**) 20 kHz groove morphology edge width, (**c**) 30 kHz groove width, (**d**) 30 kHz groove morphology edge width.

4.3. *Process Analysis of the Influence of Laser Pulse Width*

4.3.1. Process Parameter Design

According to the previous experimental design, the best cutting number of 20 times is achieved when other laser processes are unchanged, namely with the laser pulse width as the independent variable control experiment, by using a three-dimensional microscope to look at the material surface after cutting to obtain the material surface crack width under different pulse width process and by machining affected zone width data., A summary of the specific experimental parameters after data processing are shown in Table 9.

Table 9. Experimental parameters based on the pulse width.

	Crack Width/μm	Width of Machining Affected Zone/μm	Crack Width after Chip Removal/μm	Width of Groove Morphology/μm	Whether It Is Cut Off
0.15	62	379	68	34	Yes
0.2	66	365	68	29	Yes
0.25	65	376	71	32	Yes
0.3	64	355	72	23	Yes
0.35	67	368	72	26	Yes
0.4	68	366	72	28	Yes
0.45	66	364	72	24	Yes

4.3.2. Morphology Analysis before Chip Removal

By analyzing the crack width and machining affected zone width under different pulse widths, the following conclusions are obtained. (1) Cutting effect: during the pulse width experiment, the cutting times are fixed at 20 times and the materials are cut off. Changing the pulse width does not have a significant impact on this result, so the pulse width is not a laser parameter that has a significant impact on the cutting depth. (2) The pulse width of laser cutting has a certain influence on the groove width but it does not have regularity. It is certain that the groove is narrower when the pulse width is 0.3 ns, that is, the cutting quality is better. (3) The same as the influence of pulse width on the groove width, the pulse width of laser cutting has a certain influence on the width of the machining affected zone but it has no regularity. When the pulse width is 0.3 ns, the width of the machining affected zone is smaller so the laser cutting quality is better. Figure 19 shows the changing trend of relevant areas with pulse width. Figure 20 shoes the material morphology under a partial pulse width: (a) material morphology before chip removal and (b) amplification at the groove.

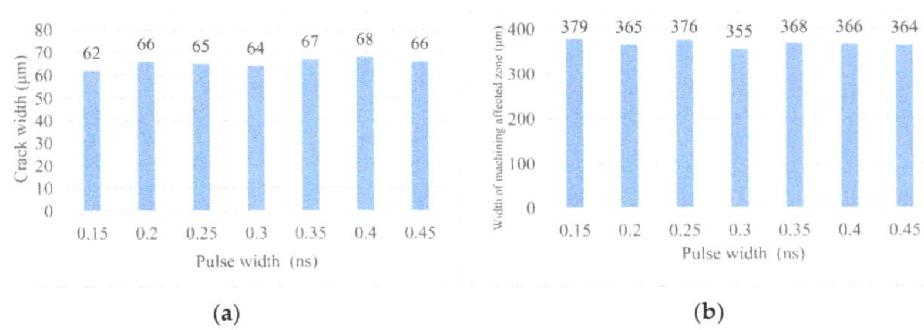

Figure 19. Change in material morphology before chip removal in the pulse width experiment. (**a**) Material crack width and (**b**) material machining the affected zone width.

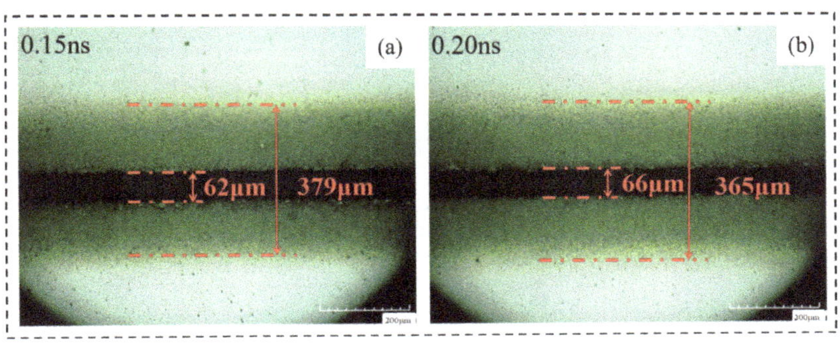

Figure 20. Material morphology at partial laser pulse width. (**a**) 0.15 ns, (**b**) 0.20 ns.

4.3.3. Morphology Analysis after Chip Removal

Through the analysis of the width of chip morphology and the width of groove morphology after chip removal with different pulse widths, the following conclusions are obtained. (1) Crack width after chip removal: The crack width after chip removal increases with the increase in pulse width because when the laser parameters change according to the trend, the laser energy absorbed on the unit length of the groove increases, the local temperature rises, and the material continues to melt, so the groove increases. (2) Groove morphology width: the surface morphology of the groove can be affected by changing the laser pulse width but there is no regularity. When the laser pulse width is 0.3 ns, the surface morphology width of the groove is small, that is, the cutting quality is good. Figure 21 shows the changing trend of relevant areas with pulse width, Figure 22. Material morphology under partial pulse width after chip removal (a) The width of the groove (b) amplification at the groove.

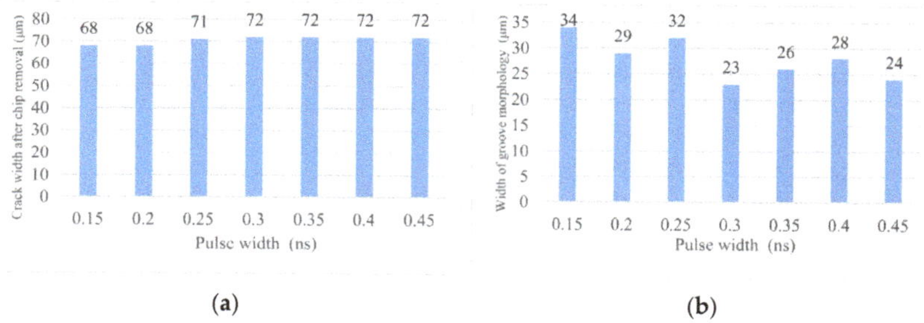

Figure 21. Change in material morphology after chip removal in the pulse width experiment. (**a**) Groove width after chip removal and (**b**) groove morphology width.

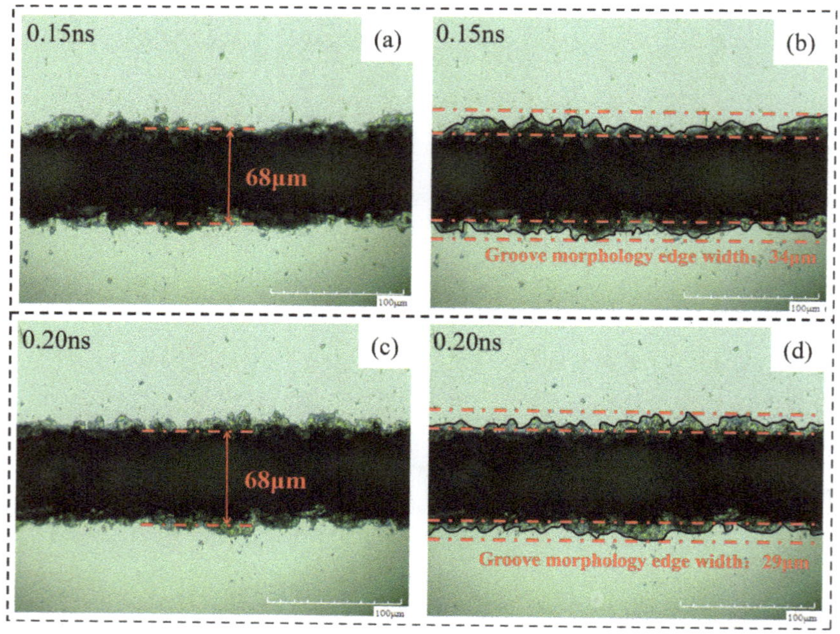

Figure 22. Material morphology after chip removal at a partial laser pulse width. (**a**) 0.15 ns groove width, (**b**) 0.15 ns groove morphology edge width, (**c**) 0.20 ns groove width, (**d**) 0.20 ns groove morphology edge width.

5. Conclusions

In view of the urgent demand of the current market for efficient wafer scribing, this paper has carried out research on the quality control and damage analysis of laser cutting of monocrystalline silicon. Through laser cutting experiments and the treatment of the material surface after cutting, the existence of thermal damage caused by laser cutting of monocrystalline silicon under the current experimental conditions and the influence of different laser parameters on the material surface quality after cutting were obtained.

(1) When laser cutting monocrystalline silicon, in addition to forming a groove at the cutting position a machining affected zone will also be formed on both sides of the groove. The size of the two zones will directly affect the cutting quality. After the material surface is wiped, the machining affected zone and the recast layer near the upper surface in the groove can be directly removed. The surface quality after wiping is basically the same as

that before cutting. In practical production, eliminating the machining affected zone by this method can improve the utilization rate of materials.

(2) With the increase in laser cutting times, the width of the material's machining affected zone and the crack width after chip removal also increase. The original crack width will also increase with the increase in cutting times when the cutting times are low but it will rapidly decrease when the cutting times exceed 80 and then slowly increase. At the same time, the change in cutting times will affect the surface morphology of the groove and the surface morphology is better when the cutting times are 20.

(3) With the decrease in the laser scanning speed, the width of the material's groove morphology decreases but the width of the material's machining affected zone and the crack width after chip removal show an increasing trend. The original crack width also shows an increasing trend when the cutting speed decreases but it will decrease rapidly when the speed decreases to 50 mm/s. The increase in laser frequency will increase the crack width and the crack width after chip removal but decrease the crack morphology and the width of the machining affected zone. The laser pulse width has a certain effect on the cutting quality but the regularity is not obvious. When the pulse width is between 0.3 ns and 0.45, the cutting quality is better, and while when the pulse width is 0.15 ns, the cutting quality is the worst.

Author Contributions: C.W. and S.Y.L. designed the whole conception and theoretical analysis. F.L. conducted the experiments and paper writing. A.Y. provided guides in the work implementation. All authors have read and agreed to the published version of the manuscript.

Funding: This work is supported by the Fundamental Research Funds for the Central Universities (2232023D-15), the China Postdoctoral Science Foundation (2022M721910), and the Shanghai Natural Science Foundation (22ZR1402400). The authors wish to record their gratitude for their generous support.

Data Availability Statement: Not applicable.

Conflicts of Interest: The authors declare no conflict of interest.

References

1. Li, W.; Jiao, Y.; Jiang, H.-Y.; Ren, Y.-H.; Ibrahim, A.M.M. Ahmed Mohamed Mahmoud Ibrahim, Investigation of mechanical force acting on the surface modified-substrate layer area during the chemical-mechanical micro-grinding of monocrystalline silicon. *Int. J. Mech. Sci.* **2022**, *228*, 107482. [CrossRef]
2. Wang, Q.; Yao, P.; Li, Y.; Jiang, L.; Xu, J.; Liang, S.; Chu, D.; He, W.; Huang, C.; Zhu, H.; et al. Inverted pyramid structure on monocrystalline silicon processed by wet etching after femtosecond laser machining in air and deionized water. *Opt. Laser Technol.* **2023**, *157*, 108647. [CrossRef]
3. Li, C.; Piao, Y.; Zhang, F.; Zhang, Y.; Hu, Y.; Wang, Y. Understand anisotropy dependence of damage evolution and material removal during nanoscratch of MgF_2 single crystals. *Int. J. Extrem. Manuf.* **2023**, *5*, 015101. [CrossRef]
4. Priolo, F.; Gregorkiewicz, T.; Galli, M.; Krauss, T. Silicon nanostructures for photonics and photovoltaics. *Nat. Nanotechnol.* **2014**, *271*, 101038. [CrossRef] [PubMed]
5. Guo, Y.; Yang, X.; Kang, J.; Li, M.; Xie, Q.; Xiao, J.; Zhang, W. Experimental investigations on the laser-assisted machining of single crystal Si for optimal machining. *Opt. Laser Technol.* **2021**, *141*, 107113. [CrossRef]
6. Wang, J.Q.; Yan, Y.D.; Li, C.; Geng, Y.Q. Material removal mechanism and subsurface characteristics of silicon 3D nanomilling. *Int. J. Mech. Sci.* **2023**, *242*, 108020. [CrossRef]
7. Zhang, W.J.; Chen, J.S.; Li, S.; Wu, Y.H.; Zhang, P.L.; Yu, Z.S.; Yue, Z.H.; Chun, Y.; Lu, H. Electronic and mechanical properties of monocrystalline silicon doped with trace content of N or P: A first-principles study. *Solid State Sci.* **2021**, *120*, 106723. [CrossRef]
8. Bian, D.; Song, E.M.; Ni, Z.F.; Qian, S.; Zhao, Y. Optimization of CMP processing parameters for Si based on response surface method. *Diam. Abras. Eng.* **2022**, *42*, 745–752.
9. Wu, C.; Li, B.; Liu, Y.; Liang, S.Y. Surface roughness modeling for grinding of silicon carbide ceramics considering co-existence of brittleness and ductility. *Int. J. Mech. Sci.* **2017**, *133*, 167–177. [CrossRef]
10. Uddin, M.S.; Seah, K.H.W.; Rahman, M.; Li, X.P.; Liu, K. Performance of single crystal diamond tools in ductile mode cutting of silicon. *J. Mater. Process. Technol.* **2006**, *185*, 24–30. [CrossRef]
11. Dong, Z.; Yan, Y.; Peng, G.; Li, C.; Geng, Y. Effects of sandwiched film thickness and cutting tool water contact angle on the processing outcomes in nanoskiving of nanowires. *Mater. Des.* **2023**, *225*, 111438. [CrossRef]
12. Wang, Y.; Zhao, B.C.; Huang, S.J.; Qian, Z.F. Study on the subsurface damage depth of monocrystalline silicon in ultrasonic vibration assisted diamond wire sawing. *Eng. Fract. Mech.* **2021**, *258*, 108077. [CrossRef]

13. Toshiro, K.; Hideo, A.; Osamu, O.; Yin, S.H.; Ren, Y.H. Challenges of future high-precision polishing methods for hard-to-process materials by the fusion of environmental control and plasma technology. *Diam. Abras. Eng.* **2022**, *42*, 637–649.
14. Costa, E.C.; Weingaertner, W.L.; Xavier, F.A. Influence of single diamond wire sawing of photovoltaic monocrystalline silicon on the feed force, surface roughness and micro-crack depth. *Mater. Sci. Semicond. Process.* **2022**, *143*, 106525. [CrossRef]
15. Costa, E.C.; Xavier, F.A.; Knoblauch, R.; Binder, C.; Weingaertner, W.L. Effect of cutting parameters on surface integrity of monocrystalline silicon sawn with an endless diamond wire saw. *Sol. Energy* **2020**, *207*, 640–650. [CrossRef]
16. Wang, B.; Melkote, S.N.; Wang, P.; Saraogi, S. Effect of speed on material removal behavior in scribing of monocrystalline silicon. *Precis. Eng.* **2020**, *66*, 315–323. [CrossRef]
17. Guo, X.Y.; Xu, Y.C.; Cao, J.F.; Zhu, J.H.; Zhao, Y.J.; Zhao, J.W.; Shi, C.Y. Laser dressing technology for micro-grooves on the surface of metal-bonded diamond wheels. *Diam. Abras. Eng.* **2022**, *42*, 364–372.
18. Naresh; Khatak, P. Laser cutting technique: A literature review. *Mater. Today Proc.* **2022**, *56*, 2484–2489. [CrossRef]
19. Li, C.; Hu, Y.; Zhang, F.; Geng, Y.; Meng, B. Molecular dynamics simulation of laser assisted grinding of GaN crystals. *Int. J. Mech. Sci.* **2023**, *239*, 107856. [CrossRef]
20. Marks, M.R.; Cheong, K.Y.; Hassan, Z. A review of laser ablation and dicing of Si wafers. *Precis. Eng.* **2022**, *73*, 377–408. [CrossRef]
21. Wu, C.; Li, J.; Hou, T.Y.; Yu, N.B.; Gao, X.J. Effect of pad and slurry on fixed abrasive polishing of gallium oxide crystal. *Diam. Abras. Eng.* **2022**, *42*, 720–727.
22. Zhao, C.Y.; Cai, Y.C.; Ding, Y.; Yang, L.J.; Wang, Z.L.; Wang, Y. Investigation on the crack fracture mode and edge quality in laser dicing of glass-anisotropic silicon double-layer wafer. *J. Mater. Process. Technol.* **2020**, *275*, 116356. [CrossRef]
23. Berhe, M.G.; Oh, H.G.; Park, S.-K.; Lee, D. Laser cutting of silicon anode for lithium-ion batteries. *J. Mater. Res. Technol.* **2022**, *16*, 322–334. [CrossRef]
24. Zhou, R.; Lin, S.; Ding, Y.; Yang, H.; Keng, K.O.Y.; Hong, M. Enhancement of laser ablation via interacting spatial double-pulse effect. *Opto-Electron. Adv.* **2018**, *1*, 180014. [CrossRef]
25. Ding, Y.; Yang, L.J.; Hong, M.H. Enhancement of pulsed laser ablation assisted with continuous wave laser irradiation. *Sci. China Phys. Mech. Astron.* **2019**, *62*, 70–78. [CrossRef]
26. Kumagai, M.; Uchiyama, N.; Ohmura, E.; Sugiura, R.; Atsumi, K.; Fukumitsu, K. Advanced dicing technology for semiconductor wafer—Stealth dicing. *Semicond. Manuf.* **2007**, *20*, 259–265. [CrossRef]
27. Huang, Y.Q.; Cao, L.; Zhang, W.L.; Hang, T.; Li, M.; Wu, Y.W. Nano-twins and stacking faults induced by picosecond laser dicing with low fluence strengthening monocrystalline silicon. *Mater. Lett.* **2023**, *334*, 133719. [CrossRef]
28. Duc, D.H.; Naoki, I.; Kazuyoshi, F. A study of near-infrared nanosecond laser ablation of silicon carbide. *Int. J. Heat Mass Transf.* **2013**, *65*, 713–718. [CrossRef]
29. Morar, N.I.; Roy, R.; Gray, S.; Nicholls, J.; Mehnen, J. Modelling the influence of laser drilled recast layer thickness on the fatigue performance of CMSX-4. *Procedia Manuf.* **2018**, *16*, 67–74. [CrossRef]
30. Yang, L.J.; Wang, Y.; Tian, Z.G.; Cai, N. YAG laser cutting soda-lime glass with controlled fracture and volumetric machining absorption. *Int. J. Mach. Tools Manuf.* **2010**, *50*, 849–859. [CrossRef]
31. Ueda, T.; Yamada, K.; Oiso, K.; Hosokawa, A. Thermal stress cleaving of brittle materials by laser beam. *CIRP Ann.* **2002**, *51*, 149–152. [CrossRef]
32. Yamada, K.; Nishioka, S.; Hosokawa, A.; Ueda, T. Cleaving process of brittle materials with pulsed YAG Laser: Thermal stress analysis in cleaving process of silicon wafer. *Jpn. Soc. Precis. Eng.* **2003**, *69*, 120–124. [CrossRef]
33. Haupt, O.; Siegel, F.; Schoonderbeek, A.; Richter, L.; Kling, R.; Ostendorf, A. Laser dicing of silicon: Comparison of ablation mechanisms with a novel technology of thermally induced stress. *Laser Microfabr. Nanomanufact.* **2008**, *3*, 135–140. [CrossRef]
34. Yamada, K.; Ueda, T.; Hosokawa, A.; Yamane, Y.; Sekiya, K. Thermal damage of silicon wafer in thermal cleaving process with pulsed laser and CW laser. *Int. Soc. Opt. Photonics* **2006**, *6107*, 61070H.
35. Shalupaev, S.; Serdyukov, A.; Mityurich, G.; Aleksiejuk, M.; Nikitjuk, Y.; Sereda, A. Modeling of mechanical influence of double-beam laser on single-crystalline silicon. *Arch. Metall. Mater.* **2013**, *58*, 1381–1385. [CrossRef]

Disclaimer/Publisher's Note: The statements, opinions and data contained in all publications are solely those of the individual author(s) and contributor(s) and not of MDPI and/or the editor(s). MDPI and/or the editor(s) disclaim responsibility for any injury to people or property resulting from any ideas, methods, instructions or products referred to in the content.

micromachines

MDPI

Article

Investigation of Cutting Force and the Material Removal Mechanism in the Ultrasonic Vibration-Assisted Scratching of 2D-SiCf/SiC Composites

Hao Lin *, Ming Zhou, Haotao Wang and Sutong Bai

School of Mechatronics Engineering, Harbin Institute of Technology, Harbin 150001, China; zhouming@hit.edu.cn (M.Z.); 18b308010@stu.hit.edu.cn (H.W.); baisutong1997@126.com (S.B.)
* Correspondence: lh927175@163.com

Abstract: Ultrasonic-assisted grinding (UAG) is widely used in the manufacture of hard and brittle materials. However, the process removal mechanism was never elucidated and its potential is yet to be fully exploited. In this paper, the mechanism of material removal is analyzed by ultrasonic-assisted scratching. Three distinct surfaces (S1, S2, and S3) were selected on the basis of the braided and laminated structure of fiber bundles. The ultrasonic-assisted scratching experiment is carried out under different conditions, and the scratching force (SF) of the tested surface will fluctuate periodically. Under the conditions of different feed speeds, depths, and ultrasonic amplitudes, the normal scratching force (SF_n) is greater than the tangential scratching force (SF_t), and the average scratching force on the three surfaces is generally S3 > S2 > S1. Among the three processing parameters, the speed has the most significant influence on the scratching force, while the scratching depth has little influence on the scratching force. Under the same conditions and surface cutting mode, the ultrasonic vibration-assisted scratching force is slightly lower than the conventional scratching force. The scratching force decreases first and then increases with the amplitude of ultrasonic vibration. Because the fiber undergoes a brittle fracture in the ultrasonic-assisted scratching process, the matrix is torn, and the surface residues are discharged in time; therefore, the surface roughness is improved.

Keywords: ultrasonic-assisted scratching; surface formation mechanism; SiCf/SiC composites; surface roughness

1. Introduction

In recent years, ceramic matrix composites (CMC), particularly carbon fiber and silicon carbide fiber-reinforced ceramic matrix composites (SiCf/SiC), received extensive attention [1,2]. A SiC ceramic matrix is a composite material with high specific strength and specific stiffness, corrosion resistance, oxidation resistance, high wear resistance, electromagnetic wave absorption, and high resistance, alongside other excellent characteristics [3,4]. However, it also has the disadvantage of being fragile, exhibiting a low fracture toughness [5]. Silicon carbide fiber is compatible with metals, resins, and ceramics, and can be used in heat-resistant, oxidation-resistant, and high-performance composite reinforcement materials. Continuous SiC fiber is used to strengthen the material, allowing the matrix's excellent properties to be maintained, and its overall toughness improved. Although the cost of the composite materials of silicon carbide fiber (compared with carbon fiber) is greater (because of the organic silicide raw materials produced by the spinning, silicification, or vapor deposition of inorganic fiber with a β-silicon carbide structure), its heat resistance and oxidation properties are better than those of carbon fiber, and the service temperature of silicon carbide fiber can reach up to 1200°. It therefore has excellent development potential in the aerospace, electronic communication, electronic machinery, petrochemical, and biomedical fields. The National Aeronautics and Space Administration (NASA) identified SiCf/SiC as the best material system for developing high-speed civil transport (HSCT) in the research results of their EPM project [6]. Subsequently, this material is used to design and prepare combustion chamber flame tubes, turbine stator cotyledons, wing front segments, thrust chambers, etc. However, this material is also characterized by brittleness and an anisotropic structure and pore characteristics, making it extremely difficult to process and mould into unique shapes and sizes [7–11]. In the process of machining SiCf/SiC ceramic composites, severe tool wear, cracks in the material matrix [12], transverse fractures [13], delamination, and fiber–matrix debonding will occur, resulting in a rough surface and a low processing efficiency, both of which directly affect the performance of the material [14,15]. Due to the material's anisotropy and hard, brittle properties, experts and engineers used various processing methods to improve its surface machining accuracy and removal efficiency. Due to the defects of delamination and burr in the traditional processing of fiber-reinforced ceramic composites, particular processing technologies, such as pulsed laser machining (PLM), high-pressure water cutting, ultrasonic machining (UM), and electrical discharge machining (EDM), are widely used in the processing of fiber-reinforced ceramic composites [16–19]. Muttamara et al. successfully drilled microholes into ceramic-based workpieces using the EDM method with an auxiliary electrode [20]. However, SiCf/SiC composites have extremely poor electrical conductivity, so the wear of the tool electrode becomes lager than with the normal machining. Hu et al. used an ultrashort pulse laser to carry out experimental research on the processing of micro through-holes and blind holes in SiCf/SiC composites; they found that it improved the processing quality and accuracy of the wall's surface, and also that the processed surface will feature a concentration of thermal stress, resulting in multiple cracks [21]. Ultrasonic machining of ceramic composite materials produces a high surface quality, but low efficiency [22,23]. Compared with the normal cutting process, rotary ultrasonic groove machining for ceramic matrix composites significantly reduces cutting force and tool wear [24,25].

In this paper, rotary ultrasonic grinding (RUG) is used to remove ceramic composites. This method can reduce the grinding force, improve the machining accuracy, and reduce the consumption of grinding tools. RUG is effective in improving the surface quality of ceramic composites and reducing tool wear [26–34]. However, the removal mechanism of the given material still needs to be fully clarified. Liu et al. studied the influence of different surface grinding forces through Cf/SiC scratching experiments [35]. Yao et al. studied the influence and removal mechanism of the scratching angle and scratching force of SiCf/SiC composites [36]. Ning et al. made a comparison between rotary ultrasonic machining of ceramic composites and conventional machining experiments. Their results show that rotary ultrasonic machining can lead to a larger ductile removal region before the successive brittle fractures and cracks [37]. Therefore, the effect of ultrasonic amplitude on surface roughness can be studied via a single-particle scratching experiment.

SiCf/SiC material is composed of a layer of orthogonal braided fiber bundles superimposed upon the matrix. Different grinding directions produce different machining effects. At present, single-particle scratching is a simple and effective method that is often used to analyze and study the removal mechanism of grinding [38–40]. Azarhoushang et al. carried out a comparative test, using ultrasonic-assisted grinding (UAG) and conventional grinding (CG) on C/C-SiC composites; they found that the grinding force was significantly reduced by 20%, and that the surface roughness was reduced by about 30% [41]. Singh R and Khamba JS et al. found that using UM in the high-precision machining of titanium alloys and other alloys can effectively prevent excessive tool wear, reduce the grinding force, and improve the grinding removal rate (GRR) [42].

In this experiment, 2D-SiCf/SiC orthogonal fiber braided stacked ceramic composites were selected as the subject. This material is more rigid and resistant to processing than Cf/SiC. Its fibers can be categorized by three orthogonal directions: transverse fibers, longitudinal fibers, and normal fibers. The effects of grinding in different directions under ultrasonic vibration on the grinding force and machined surface roughness were analyzed. The influence of ultrasonic vibration grinding on three typical grinding surfaces was studied using a Vickers indenter via a single-particle scratching experiment. Different scratch speeds and depth factors were used to study SiCf/SiC composites, obtain the scratching force (S.F.), and deeply analyze images of the composites' surface morphology. Finally, the effects of ultrasonic vibration scratching and traditional scratching on SiCf/SiC's material removal rate and surface roughness are compared. This study may provide technical support for improving the machined surface quality of woven fiber-reinforced ceramic matrix composites.

2. Material and Rotating Ultrasonic Machining (RUM) Platform

2.1. Material Preparation

In the experiment, a SiCf/SiC composite with a two-dimensional orthogonal braided structure was used as the material. Figure 1a shows a scanning electron microscope (SEM) image, and Figure 1b shows an illustration, produced using software, of the braided structure of the SiC fibers. A physical view of the material before processing is shown in Figure 1c. The microstructure of the composite material is shown in Figure 2. The thickness of the orthogonal SiC fiber bundle layer is about 300 μm~500 μm. In this layer, the width of the SiC fiber bundles is about 700 μm, and the arrangement directions are 0° and 90°, respectively. In the fetal layer of SiC fibers, the SiC fibers are composed of SiC fiber bundles 14 μm in diameter, which are irregularly interwoven to form a net-like structure. It can be seen from the image that some SiC fiber bundles are similar to disordered networks, and orderly orthogonal distribution can also be seen in some areas. This is mainly caused by the anisotropic characteristics of the SiCf/SiC material and the internal heterogeneous structure. Therefore, for a single-particle scratching test, irregular fiber network layers should be avoided, and the orthogonal SiC fiber layer should instead be selected as the scratching surface. A simple polishing treatment is needed before the test can proceed. The properties in Table 1 were obtained by testing the material's physical and mechanical properties and reviewing the resulting data. The sample size of the scratching test material is 15 × 12 × 3 mm. In the polishing process, 40 μm abrasive particles were first selected for grinding (for 2 min); then, 15 μm abrasive particles were chosen for grinding (for 5 min); and finally, 1 μm abrasive particles with suspension were used to grind the surface for 10 min. In this way, the surface can meet the requirements of the test.

Table 1. Material properties of SiCf/SiC [43].

Parameters	Density (g/cm^3)	Fiber Volume Fraction (%)	E-Modulus (GPa)	Tensile Strength (Mpa)	Flexural Strength (Mpa)	Compression Strength (Mpa)
Values	2.5 ± 0.03	23 ± 3	220 ± 10	271.4 ± 29.0	500 ± 50	454.0 ± 18.2

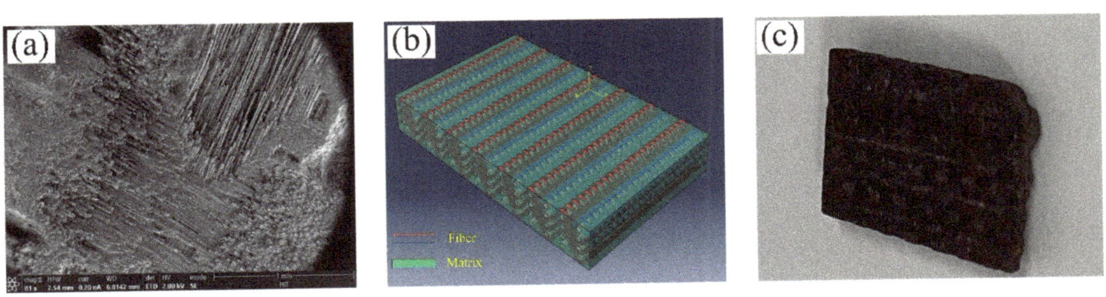

Figure 1. Schematics of the structure of the SiCf/SiC workpiece material: (**a**) SEM image, (**b**) plane woven structure of the SiC fibers and matrix, and (**c**) images of the SiCf/SiC composite.

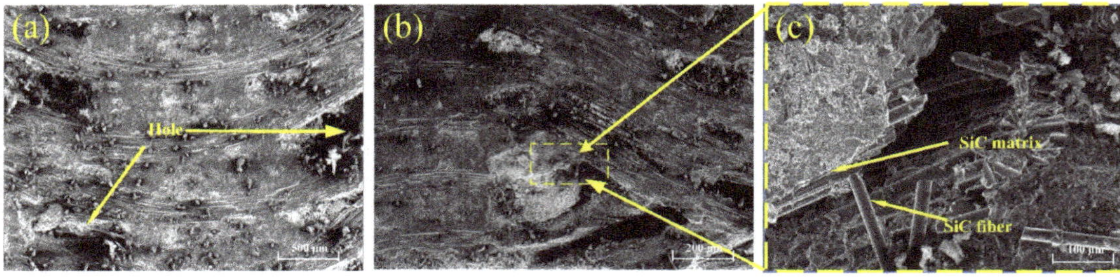

Figure 2. SiCf/SiC composite structure: (**a**) SiC fiber braided mesh layer, (**b**) section diagram of SiCf/SiC composites, and (**c**) ceramic matrix.

2.2. Experimental Design and Conditions

The CNC five-axis machining center (Ultrasonic 70-5 linear, DMG MORI, Bielefeld, Germany) was used in the experiment. The USG2000 ultrasonic vibration system equipped with the machining tool had a maximum ultrasonic vibration frequency of 30 kHz. The vibration mode of the tool is such that the spindle moves up and down, and the axial vibration is only one-dimensional (perpendicular to the material surface). A standard Vickers diamond indenter (THV-6) was used for variable cutting depth testing, as shown in Figure 3. The diagonal and relative angles of the indenter are 136°. The Vickers diamond indenter was installed on the HSK63 shank, as shown in Figure 4. The test material was a 2D-SiCf/SiC ceramic composite, and the workpiece size was 15 × 12 × 3 mm. Before the test, the material only needed simple treatment. The test material was fixed on the machine tool with a particular fixture, and the machine tool program was controlled to maintain a certain angle. Thus, the experimental results were obtained by cutting the indenter on the surface of the material. Then, the plate was screwed firmly onto the K9527B dynamometer (Kistler, Winterthur, Switzerland). Finally, the dynamometer was fixed on the machine tool table with a fixture, as shown in Figure 4. The surface features were measured using a laser confocal microscope (LEXT OLS3000, OCPNY, Tokyo, Japan) to obtain the roughness.

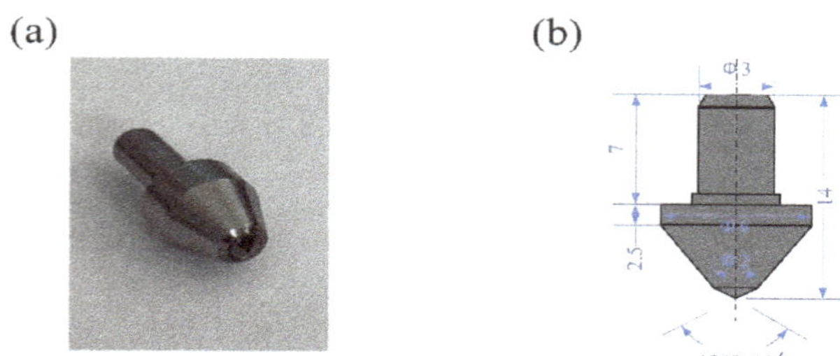

Figure 3. Pictures and detailed parameters of the Vickers diamond indenter. (**a**) Physical image of indenter, and local enlarged image; and (**b**) experimental indenter size parameters.

The microhardness was tested with a CLEMEX ST-2000 (Clemex, Technologies, Montreal, QC, Canada) digital display. The test instrument can realize the automatic loading and unloading process, as it is equipped with an automated electric loading platform. The working accuracy of the X and Y-axis can reach 0.5 μm, while the autofocus function can reach 0.1 μm on the Z-axis. The effects of three directions on the surface quality and cutting force of SiCf/SiC ceramic composites with the same amplitude were studied. The workpiece is clamped in a triaxial piezoelectric dynamometer (Kistler 9257B, Kistler, Winterthur, Switzerland) with a sampling frequency of 7 kHz. On the surface of the workpiece, the two components of the experimental scratching forces (SFs) are F_x and F_z. Where $F_n = -F_z$ (normal SF) and $F_t = -F_x$ (tangential SF). Where S_v is the scratching feed speed, and S_d is the scratching depth, as shown in Figure 5a. By rotating the table of the DMG machining center around the axis by a certain angle, the testing of the variable cutting depth can be accomplished, as shown in Figure 5b. Then, the conventional scratching and ultrasonic-assisted scratching were observed using scanning electron microscopy (SEM).

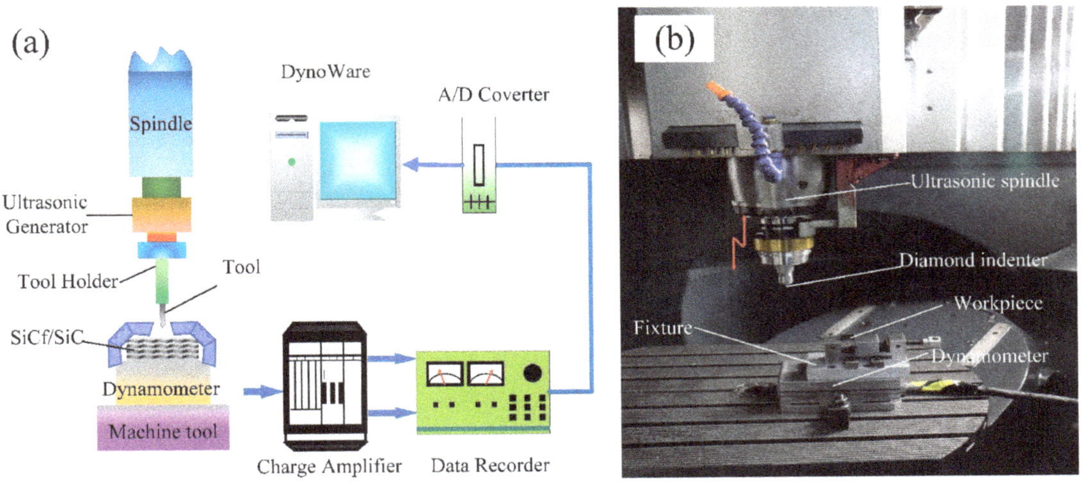

Figure 4. System set-up for scratching tests: (**a**) schematic diagram of the overall structure; and (**b**) photo of the experimental device.

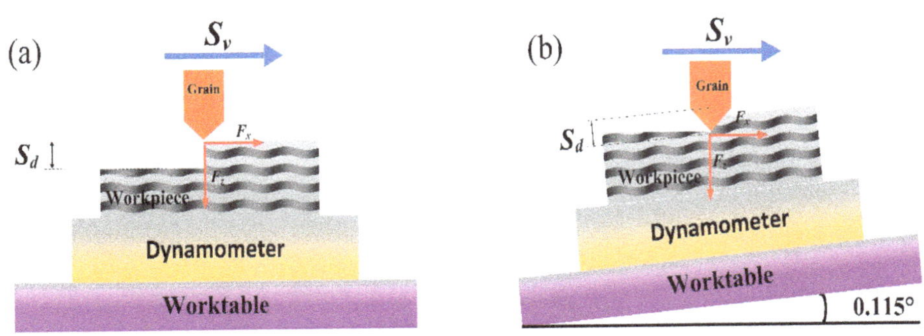

Figure 5. Single-grain scratching experiment. (**a**) single grain conventional scratching and (**b**) single grain scratching with a gradual depth.

2.3. Scratching Experiment

Due to the anisotropy and heterogeneity of the material, the material can be divided into three typical planes, S1, S2, and S3, as shown in Figure 6. According to the designed direction and selected cutting surface in the figure, S1 represents the vertical plane composed of 0° fiber bundles and 90° fiber bundles, S2 is composed of cross-fibers and 0° fiber bundles, and S3 is mainly composed of 90° fiber bundles and cross-fibers. Subsequently, ultrasonic scratching tests were carried out on the three vertical surfaces, S1, S2, and S3, respectively. The parameters used in the scratching process are shown in Table 2. For each group of test data, the scratching mode, as shown in Figure 7, was adopted for these three surfaces, and each group of test data was collected five times; afterwards, the mean scratching force was calculated. After the test, each workpiece was placed under the scanning electron microscope (SEM) to observe the specific situation after scratching. Finally, the roughness was measured using a laser confocal microscope (LEXT OLS3000) instrument.

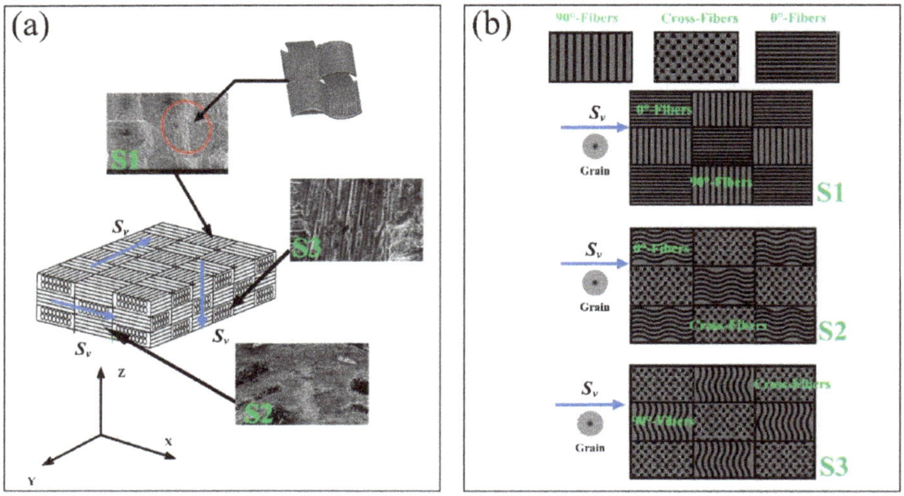

Figure 6. Definition of typical surfaces and scratching directions of composites: (**a**) 3D braided structure and cutting directions of SiCf/SiC materials; and (**b**) definitions of the scratched surfaces of S1, S2, and S3.

Table 2. Scratching parameters for single-grain scratching dates.

Surfaces	Parameters	Values	Ultrasonic Amplitude (μm)
S1, S2, S3	Feeding speed S_v (mm/min) Scratching depth S_d (μm)	200, 400, 600 20, 30, 40, 50	3

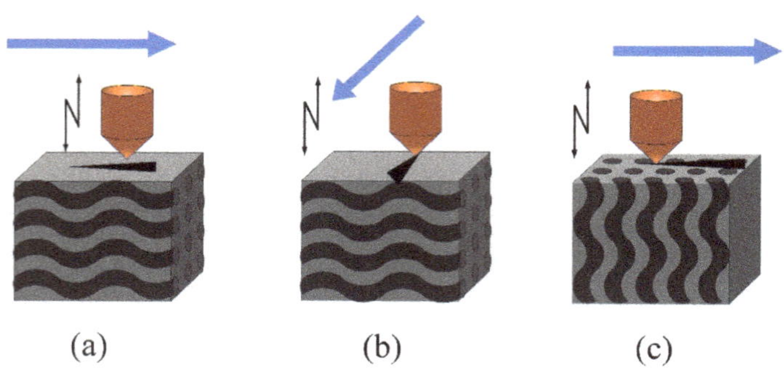

Figure 7. Definition of the scratching mode and the given direction of vibration: (**a**) longitudinal scratching; (**b**) transverse scratching; and (**c**) normal scratching.

3. Results and Discussion

3.1. Scratching Force and Surface Morphology of Three Typical Braided Surfaces

In the SEM images of the two kinds of machined microsurface morphology in Figure 8, it can be seen that the morphology of the fiber fracture surface is irregular, and the ultrasonic vibration-assisted scratching (UVAS) method is slightly better than the conventional scratching (CS) method. The removal patterns of composites processed in both ways are similar. The overall performance of the fiber is one of brittle fracture removal, mainly in the form of collective crushing, fiber fracture, the fiber–matrix interface being off-site, peeling, and so on. The strength of the matrix is greater than that of the fiber, and the interface adhesion between the fiber and the matrix is relatively weak, meaning that many pieces fall away. Therefore, the mode of the SiCf/SiC removal is brittle.

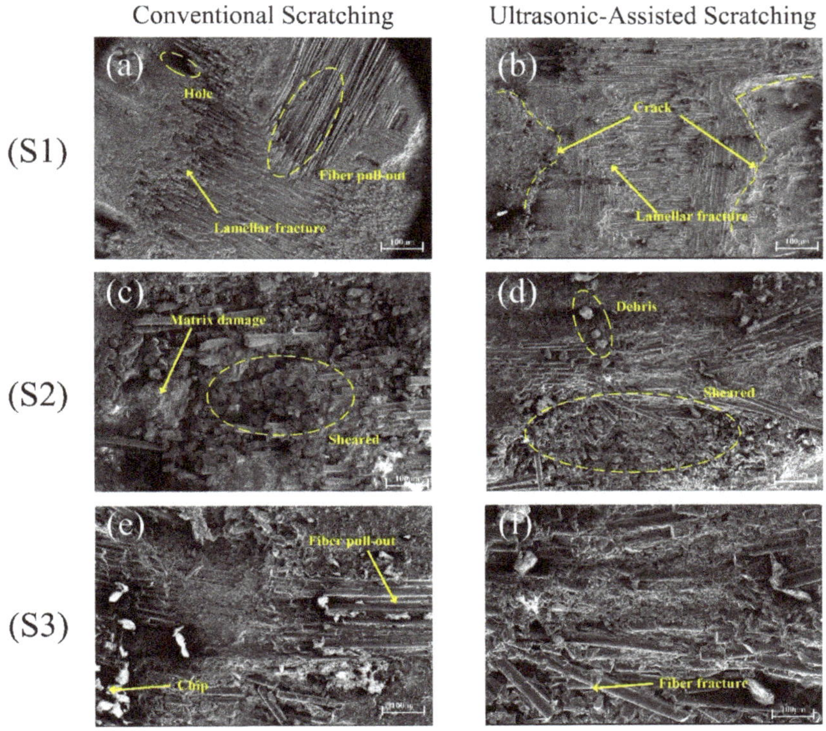

Figure 8. Morphologies of the scratched surfaces of S1 (**a**,**b**), S2 (**c**,**d**), and S3 (**e**,**f**).

The force data of S1, S2, and S3 on three surfaces were collected by a 9257B dynamometer, and ultrasonic-assisted machining was carried out using a DMG-ultrasonic70-5-linear machining tool. It can be seen from the results that the scratching force Fz on the three surfaces and the plane forces F_x and F_y change periodically and uniformly. F_x and F_y constitute the tangential force, and the normal scratching force is larger than the tangential forces. The weaving form of the surface of the material changes periodically, and the corresponding scratching force is also different; the surface scratching force is measured by averaging the scratching force when calculating the scratching force. Changes in the scratching speed and depth influence the SF, as shown in Figures 9 and 10, respectively.

It can be seen from Figure 9a,b that with the increase in velocity, the scratching force on each scratching surface increases significantly and obviously; this may be because the selected velocity values vary wildly. With the increase in scratching depth from 20 μm to 50 μm, the scratch force on the three surfaces increases evenly, but not significantly. An increase in scratch depth directly leads to an improvement in the material removal rate. However, from Figures 9 and 10, it is clear that the surface scratching force of S3 is greater than the other two surface forces. Under the influence of the anisotropy of the woven and laminated structures of the composite, the normal and tangential scratching forces of the different plane engravings follow the size order S3 > S2 > S1. As shown in Figure 9, the tangential scratching force (SF_t) of the three planes is less than the normal scratching force (SF_n), which is about half of SF_n; with the increase in feed speed, both SF_n and SF_t increase. Because the single particle increases in unit time with the increase in feed speed, the processing efficiency is improved, and the amount of removal achieved per unit of time is increased; therefore, the SFs increases. In Figure 10, it is shown that the SFs increase with increasing depth. The SF_t and SF_n in the S3 plane change significantly more than the forces in the two planes of S1 and S2, and these differences are mainly attributed to the lamination factor. In the S3 plane, the fibers are characterized by X-Y fiber weaving; they are then hot-melt laminated in the Z direction to form a composite material. The two-dimensional structural properties significantly reduce anisotropy, so the shear strength and tensile strength in the lamination plane are the worst combination therein [32]. At the same time, the debonding and fiber peeling between the layers are very tolerant of the fiber/matrix of the plane, so the scratching force of the laminate plane is the lowest.

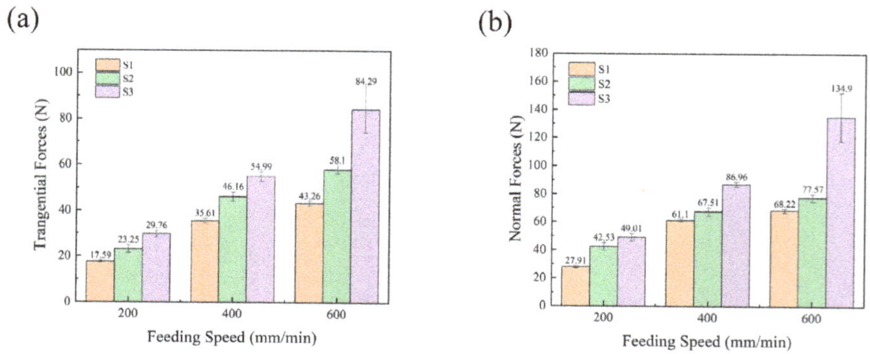

Figure 9. The effects of scratching speed on components of SF, under ultrasonic vibration machining, (A_m = 3 μm, S_d = 20 μm): (**a**) tangential SF and (**b**) normal SF.

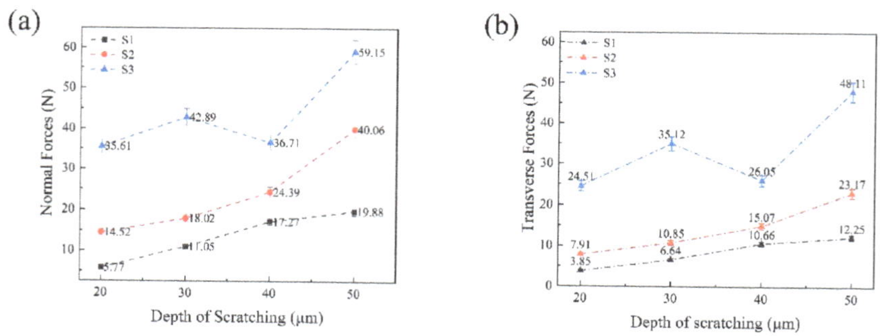

Figure 10. The effects of scratching depth on components of SF under ultrasonic vibration machining, (A_m = 3 μm, S_v = 200 mm/min): (**a**) normal SF, and (**b**) tangential SF.

Two scratching methods were used according to the scratching direction shown in Figure 11. According to the DMG operation manual and display interface, the accuracy of the rotary axis is 0.001°. The turntable was rotated by 0.115° to achieve a 1 mm scratching at a depth of 2 μm. In the beginning, when the blade tip is drawn, the depth of the cutting is shallow, the damage to the surface of the composite material is minimal, and the difference in the comparative cutting force is not apparent. However, it can be seen that the primary forms of failure are matrix fracturing, fiber debonding, peeling, and fiber fracturing. The ceramic fiber material S1 has a transverse and longitudinal fiber bundle braided layer on its surface. Therefore, parallel and identical woven forms were chosen before scratching observations began.

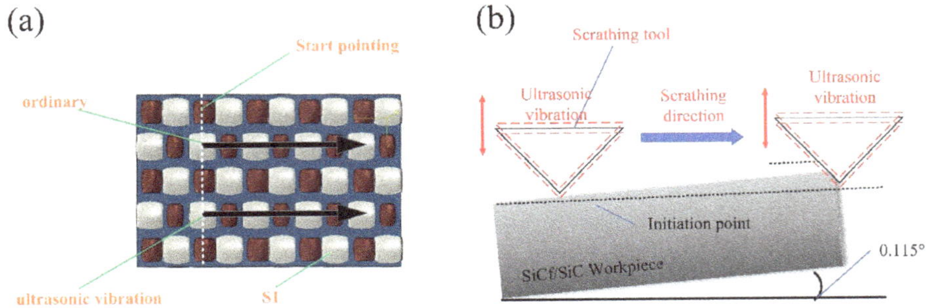

Figure 11. Diagram of fiber and scratch directions: (**a**) is the top view and (**b**) is the right view.

When the indenter is used for normal scratching, the matrix is slightly deformed in the initial stage, after which many long fibers come off and the matrix is seriously fractured. The microscopic surface morphology is shown in Figure 12a. When the material is scratched using ultrasonic vibration, there is no peeling phenomenon over a long distance, and the roughness of the matrix is smaller than that achieved with ordinary crushing, as can be seen in Figure 12b. The fiber's breaking time is short, resulting in a low fiber debonding rate and a smooth break removal.

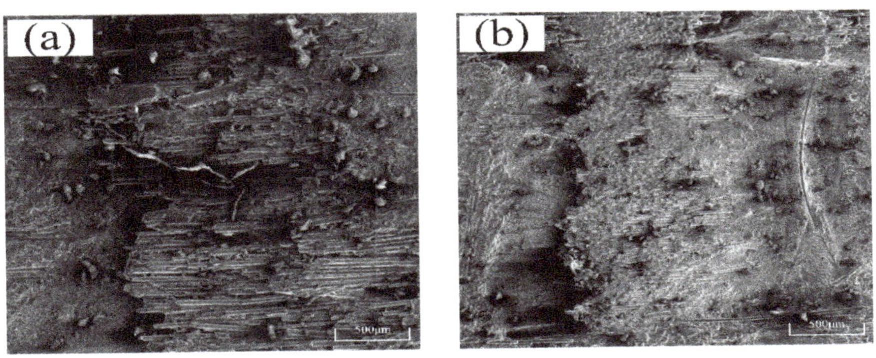

Figure 12. Scanning electron morphologies of the scratched surface of S1, S_v = 10 mm/s, θ = 0.115°: (**a**) A_m = 0 μm, and (**b**) A_m = 3 μm (A_m represents amplitude).

According to the analysis in Figures 12 and 13, the brittleness of the composite matrix is greater than that of the SiC fiber in the CS process, and the SiC matrix begins to crack due to scratching friction. The fracture energy is dispersed as the crack extends to the interface between the matrix and fiber. The fracture is temporarily blocked and begins to expand at the boundary with the increase in shear and extrusion pressures. In the CS process, only SiC fibers are crushed into fragments, and longer fibers may be separated from the matrix. In the process of UVAS, the axial impact is added based on CS. The Z-axis is attached, and ultrasonic vibration can cut the fiber into smaller pieces that are easier to remove. As the tool moves up and down, the tool cools effectively, so the tip holds well. At the same time, the up and down vibration causes the slag and fibers to excrete more easily; thus, the surface quality improves.

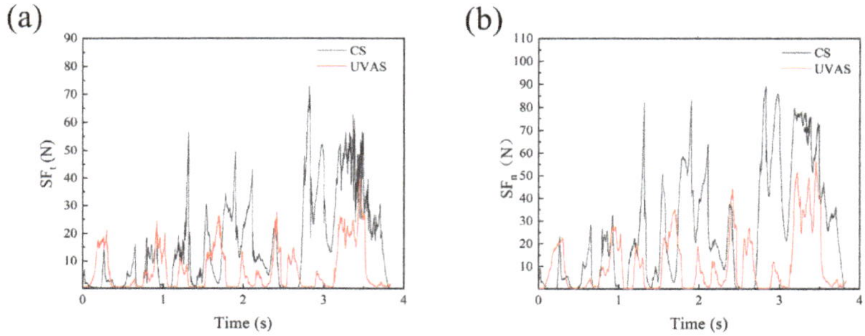

Figure 13. Variation in the measured scratching force (SF) with the cutting depth on the surface of S1: (**a**) conventional scratching and (**b**) ultrasonic vibration-assisted scratching.

3.2. Ultrasonic Scratch Morphology Observation and Removal Mechanism Analysis

SiCf/SiC composites are composed of the SiC matrix and SiC fiber-braiding stack inside the complex; the material's removal occurs in two stages. The first is when the brittle matrix failure occurs. The internal SiC fiber is debonded and stripped; a schematic diagram of this is shown in Figure 14. When the diamond touches the material surface, plastic deformation occurs due to stress. With the increase in the diamond's entry depth, the stress on the surface gradually increases, and the surface begins to feature brittle fractures and internal, transverse, and longitudinal radial cracks [44,45].

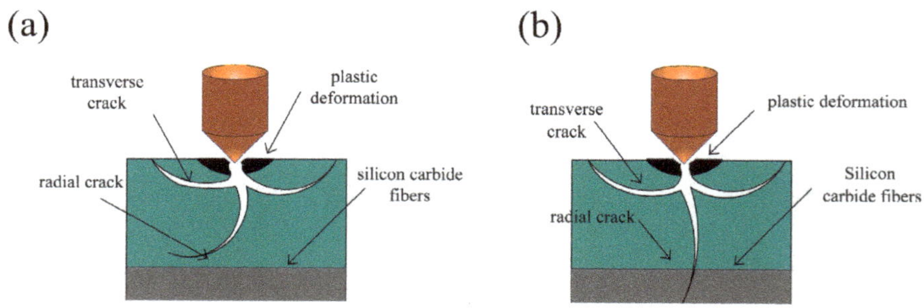

Figure 14. Single-particle grinding fracture model of SiCf/SiC composites: (**a**) crack deflection and (**b**) fiber breakage.

The SiCf/SiC composite comprises a SiC matrix phase and SiC fiber phase. The removal mechanism of the composite is different from that of ordinary materials. The SiCf/SiC composite features the toughening effect of SiC fiber, in addition to the brittleness and fracturability of the matrix. A single fracture diagram of the SiCf/SiC composite is shown in Figure 14 below. When the abrasive particles are in contact with the surface of the SiCf/SiC composite, the material first produces a small deformation due to extrusion and shear stress. With the increase in abrasive particle pressure, the cutting depth also increases, and the surface stress of the abrasive particles on the SiCf/SiC material increases. When the stress increases to a particular value, the material begins to show transverse cracks and longitudinal radial cracks. With the continuous action of the abrasive particles, the cracks in both directions extend further, and the degree of damage is related to the material's fracture toughness. When the stress of abrasive particles on the material is less than the fracture toughness, the longitudinal radial cracks' propagation direction changes and deflects in the matrix, as shown in Figure 14a. Suppose that the stress of the abrasive particles on the material continues to increase and is greater than the fracture toughness; in this case, the crack will expand to the SiC fiber, and the fiber will suffer from brittle fracture, peeling off from the matrix, and other forms of damage, as shown in Figure 14b.

According to the above experimental data, each surface will undergo the following stages: (1) When the extrusion pressure is small, the surface undergoes slight plastic deformation. (2) Due to the brittleness of the SiC matrix being greater than that of the SiC fiber, the SiC composite cracks and the stress increases. (3) When the pressure increases, the interface cracks between fiber and matrix continue to expand, and the phenomena of fiber and matrix peeling, shedding, and tearing occur. (4) The silicon carbide fiber breaks directly into pieces and falls from the material surface.

In Figure 15, it can be observed that ultrasonic-assisted vibration grinding can scratch the SiC fiber in different directions, and the results are entirely different. In Figure 15a,b, in the horizontal transverse direction, we can see that the average score of the fibers due to tangential force is far more considerable than the SiC fiber and substrate, hence the long fiber breaks off after ultrasonic vibration machining is added; this is due to the up and down shaking force, which cuts into the direction of short fibers, therefore producing a small loss. In Figure 15c,d, with horizontal tangential scratching, the fibers are fractured due to shear and extrusion, and the fibers on both sides of the fracture spill off on to the matrix, but are relatively uneven. However, the ultrasonic effect means that the press head used for processing on both sides of the crack can cause the fiber to be more evenly cut. As shown in Figure 15e,f, there are many fiber holes in SiCf/SiC ceramic composites, and the fiber fracture toughness cannot withstand the large cracks caused by axial force after the ordinary cutting of the SiC matrix. However, when the ultrasonic-assisted fraction vibrates upward and downward, the matrix can form small fault blocks, and the chips can discharge in time without secondary damage. The fiber toughness can retain a small matrix block on the workpiece without causing a large area of surface breakage, thereby improving the surface roughness.

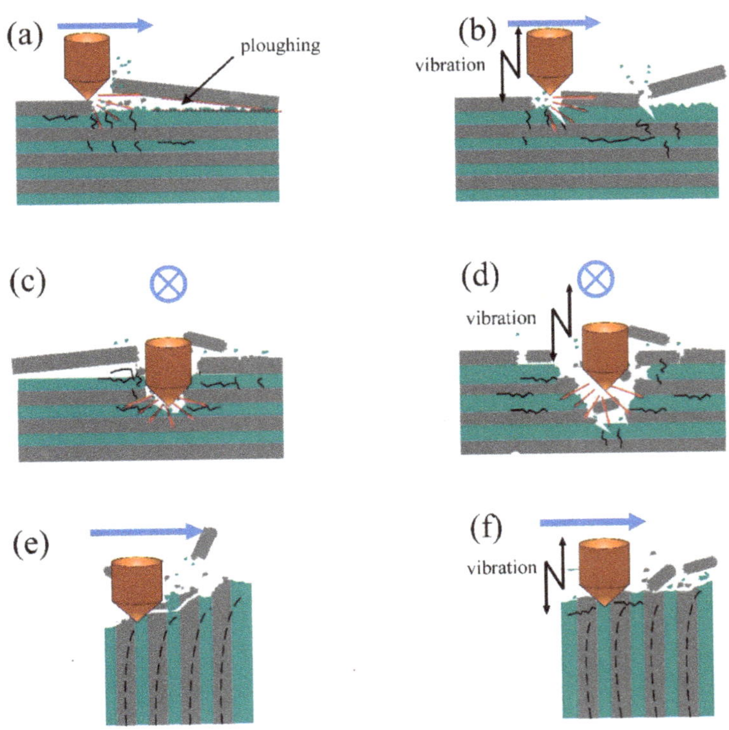

Figure 15. Influence of ultrasonic vibration on material removal mode: (**a**,**c**,**e**) represent ordinary scratching; (**b**,**d**,**f**) represent ultrasonic vibration-assisted scratching.

3.3. Study on the Influence of Ultrasonic Vibration on Surface Roughness

The difference in roughness cannot be fully seen simply by viewing the 2D morphology. The 3D characterization method can illustrate the difference between the height of each position and can accurately and comprehensively reflect the different microstructures. By establishing the function relation, the average arithmetic height S_a was calculated using a computer. S_z expressed the surface topography of the composite, and the surface roughness was objectively expressed as S_q, where S_a represents the mean value of an absolute value within a limited region.

$$S_a = \frac{1}{MN}\sum_{j=1}^{N}\sum_{i=1}^{M}|\eta(xi,yj)| \tag{1}$$

x_i and y_j are the distances from the x-direction and y-direction points on the outline to the reference line, respectively. M and N, respectively, are included in order to adopt the x-direction and y-direction measurement points in the region; η is the measurement accuracy coefficient.

The S_q (root mean square height) is equal to the standard deviation of the height distribution, and the mean square value is also known as the validity in physics, which is a convenient statistical method.

$$S_q = \frac{1}{MN}\sqrt{\sum_{j=1}^{N}\sum_{i=1}^{M}|\eta^2(xi,yj)|} \tag{2}$$

Figure 16a shows that the tangential force is larger than the scratching force as a whole; both decrease first and then increase with the increase in amplitude. When the amplitude is 2.5 μm, the normal and tangential forces are the smallest. The white light interferometer was used to measure the sampling surface of each scratch; the obtained data were statistically averaged and plotted into a chart, as shown in Figure 16b, for a comparative analysis of roughness. It can be seen that the roughness of the ultrasonic vibration-scratched surface is significantly greater than that of the ordinary scratched surface, and the S_q roughness is increased by 34~51%.

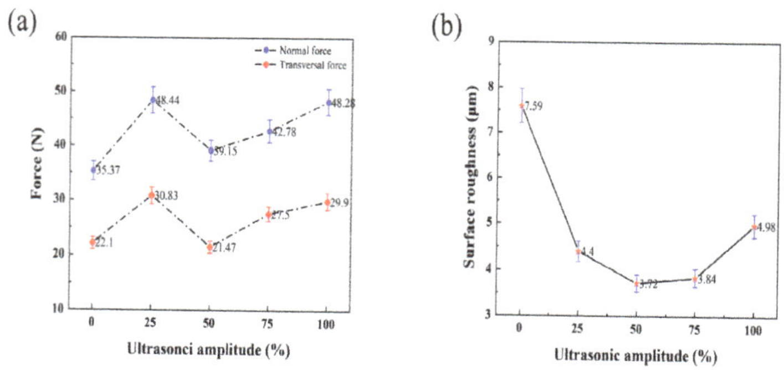

Figure 16. Variation in surface scratching force and surface roughness under different amplitudes (S_v = 200 mm/min, A_p = 20 μm, and A_m = 5 μm). (**a**) The tangential force is larger than the scratching force as a whole; both decrease first and then increase with the increase in amplitude. (**b**) The white light interferometer was used to measure the sampling surface of each scratch.

In the process of face grinding, due to ultrasonic vibration, the diamond material's abrasive grains maintain discontinuous contact with the workpiece material and impact the workpiece surface at a high frequency. Since the speed of impact is extremely fast and much greater than the speed of feeding, the high-frequency result is the primary removal mode in the face grinding process. The ratio of normal grinding force to tangential grinding force (F_n/F_t) is used to evaluate the grinding performance of the grinding wheel. The greater the grinding force, the worse the performance of the grinding wheel. As shown in Figure 17b, the quality is worst with a surface grinding amplitude of 0 μm. As the amplitude increases, the surface quality deteriorates, a trend similar to the grinding force ratio presented in Figure 17a. Notably, when the amplitude is 1.25 μm, the surface quality is the best, and the grinding force ratio is the smallest.

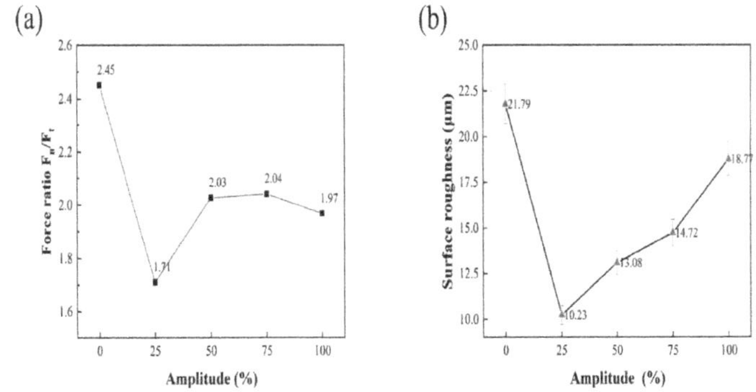

Figure 17. Effect of ultrasonic amplitude on grinding force ratio and surface roughness (S = 12,000 r/min, S_v = 25 mm/min, A_p = 25 μm, and A_m = 5 μm). (**a**) The grinding force ratio, The greater the grinding force, the worse the performance of the grinding wheel. (**b**) The ratio of normal grinding force to tangential grinding force (Fn/Ft) is used to evalu-ate the grinding performance of the grinding wheel.

Through ultrasound-assisted scratching and grinding experiments, it is not difficult to see the influence that these processes have on reducing the anisotropic processing of SiCf/SiC composites to a certain extent. SiC fibers are cut into multi-truncated fibers by ultrasonic impact and are removed by short fibers, which reduces the breaking and peeling of SiC fibers in conventional scratching. With the amplitude increase in the marking experiment, the surface scratching force first increased and then decreased. The roughness measurements gradually increased, but all were lower than those of ordinary machined surfaces. The end face grinding experiment further verifies that ultrasonic amplitude has a significant influence on the improvement of surface roughness. Said ultrasonic amplitude reduces the contact time between the abrasive particles and the material; thus, the grinding force is significantly reduced, and the chips can be discharged in time. The ultrasound-assisted impact promotes some small cracks in the composite, making the material easy to remove. Both experiments verify that the surface roughness and quality of composite materials are improved by ultrasonic-assisted processing.

4. Conclusions

This work mainly studies the grinding mechanism of 2D-SiCf/SiC using ultrasonic vibration grinding. The characteristics of scratching force and the surface morphology of three typical braided surfaces were studied. The following conclusions were obtained:

Observing the contrast between ultrasonic scratching and ordinary scratching, conventional scratching prompts severe fiber peeling and debonding, as well as larger matrix fracture cracks. In ultrasonic vibration mode, the fracture fiber becomes shorter, the chip becomes smaller, the crack fracture becomes smaller, and the overall surface smoothness is relatively better.

When three planes are scratched using ultrasonic vibration, the changing trend in scratching force is roughly the same. With an increase in the scratching speed, the normal and tangential scratching forces also increase. With an increase in scratching depth, the normal and tangential scratching forces show little change. However, with a change in the direction of the indentation head as it contacts the fiber braid, the scratching force is significantly different. The scratching force can be divided into S1, S2, and S3. The scratching forces on the three surfaces generally follow the order S3 > S2 > S1.

With the same feed rate and grinding depth, different amplitudes can reduce the grinding interaction time between abrasive particles and materials. Their combined action reduces the force and signs of subsurface damage such as matrix cracks and fiber debonding. The surface fibers of the material are subject to amplitude vibration; truncated short fibers also become neater and chips are easily discharged, so this form of processing is more conducive to a high-quality material surface.

Author Contributions: Methodology, H.W. and S.B.; Software, H.L.; Validation, H.L.; Formal analysis, H.L.; Investigation, H.L.; Resources, M.Z.; Data curation, H.L.; Writing—original draft, H.L.; Writing—review & editing, H.L.; Supervision, M.Z., H.W. and S.B.; Project administration, M.Z.; Funding acquisition, M.Z. All authors have read and agreed to the published version of the manuscript.

Funding: This work was supported by the National Natural Science Foundation of China (Grant No. 51975153).

Data Availability Statement: Data available on request from the authors.

Acknowledgments: We thank the Department of Aeronautics and Astronautics of the Harbin University of Technology for the help of their manufacturing platform and the technical support of the staff.

Conflicts of Interest: The authors declare no conflict of interest.

References

1. Diaz, O.G.; Luna, G.G.; Liao, Z.R.; Axinte, D. The new challenges of machining Ceramic Matrix Composites (CMCs): Review of surface integrity. *Int. J. Mach. Tools Manuf.* **2019**, *139*, 24–36. [CrossRef]
2. Diaz, O.G.; Axinte, D.A. Towards understanding the cutting and fracture mechanism in Ceramic Matrix Composites. *Int. J. Mach. Tools Manuf.* **2017**, *118*, 12–25. [CrossRef]
3. Naslain, R.R. SiC-matrix composites: Nonbrittle ceramics for thermo-structural application. *Int. J. Appl. Ceram. Technol.* **2005**, *2*, 75–84. [CrossRef]
4. Ma, Q.S.; Liu, H.T.; Pan, Y.; Liu, W.D.; Chen, Z.H. Research Progress on the Application of C/SiC Composites in Scramjet. *J. Inorg. Mater.* **2013**, *28*, 247–255. [CrossRef]
5. Teti, R. Machining of Composite Materials. *CIRP Ann.* **2002**, *51*, 611–634. [CrossRef]
6. DiCarlo, J.A. Advances in SiC/SiC Composites for Aero-Propulsion. *Ceram. Matrix Compos. Mater. Model. Technol.* **2014**, 217–235. [CrossRef]
7. Yin, J.F.; Xu, J.H.; Ding, W.F.; Su, H.H. Effects of grinding speed on the material removal mechanism in single grain grinding of SiCf/SiC ceramic matrix composite. *Ceram. Int.* **2021**, *47*, 12795–12802. [CrossRef]
8. Gong, Y.D.; Qu, S.S.; Yang, Y.Y.; Liang, C.Y.; Li, P.F.; She, Y.B. Some observations in grinding SiC and silicon carbide ceramic matrix composite material. *Int. J. Adv. Manuf. Technol.* **2019**, *103*, 3175–3186. [CrossRef]
9. Zhai, C.; Xu, J.; Hou, Y.; Sun, G.; Zhao, B.; Yu, H. Effect of fiber orientation on surface characteristics of C/SiC composites by laser-assisted machining. *Ceram. Int.* **2022**, *48*, 6402–6413. [CrossRef]
10. Ran, Y.C.; Kang, R.K.; Dong, Z.G.; Jin, Z.J.; Bao, Y. Ultrasonic assisted grinding force model considering anisotropy of SiCf/SiC composites. *Int. J. Mech. Sci.* **2023**, *250*, 108311. [CrossRef]
11. Lu, S.J.; Zhang, J.J.; Li, Z.Q.; Zhang, J.G.; Wang, X.H.; Hartmaier, A.; Xu, J.F.; Yan, Y.D.; Sun, T. Cutting path-dependent machinability of SiCp/Al composite under multi-step ultra-precision diamond cutting. *Chin. J. Aeronaut.* **2021**, *34*, 241–252. [CrossRef]
12. Diaz, O.G.; Axinte, D.A.; Butler-Smith, P.; Novovic, D. On understanding the microstructure of SiC/SiC Ceramic Matrix Composites (CMCs) after a material removal process. *Mater. Sci. Eng. A* **2019**, *743*, 1–11. [CrossRef]
13. Liu, Q.; Huang, G.; Xu, X.; Fang, C.; Cui, C. Influence of grinding fiber angles on grinding of the 2D–Cf/C–SiC composites. *Ceram. Int.* **2018**, *44*, 12774–12782. [CrossRef]
14. Du, J.; Ming, W.; Ma, J.; He, W.; Cao, Y.; Li, X.; Liu, K. New observations of the fiber orientations effect on machinability in grinding of C/SiC ceramic matrix composite. *Ceram. Int.* **2018**, *44*, 13916–13928. [CrossRef]
15. Xing, Y.; Deng, J.; Zhang, G.; Wu, Z.; Wu, F. Assessment in drilling of C/C-SiC composites using brazed diamond drills. *J. Manuf. Process.* **2017**, *26*, 31–43. [CrossRef]
16. Tawakoli, T.; Azarhoushang, B. Intermittent grinding of ceramic matrix composites (CMCs) utilizing a developed segmented wheel. *Int. J. Mach. Tools Manuf.* **2011**, *51*, 112–119. [CrossRef]
17. Yang, M.; Song, X.; Lin, B.; Yang, Y.; Peng, W. Study on Acoustic Emission Characteristics of Fiber Reinforced Ceramic Matrix Composites. *Mater. Sci. Forum* **2013**, *770*, 202–206. [CrossRef]

18. Cao, X.Y.; Lin, B.; Wang, Y.; Wang, S.L. Some Observations in Grinding Surface Quality of FRCMC. *Mater. Sci. Forum* **2013**, *770*, 198–201. [CrossRef]
19. Li, C.; Hu, Y.; Zhang, F.; Geng, Y.; Meng, B. Molecular dynamics simulation of laser assisted grinding of GaN crystals. *Int. J. Mech. Sci.* **2023**, *239*, 107856. [CrossRef]
20. Muttamara, A.; Fukuzawa, Y.; Mohri, N.; Tani, T. Probability of precision micro-machining of insulating Si3N4 ceramics by EDM. *J. Mater. Process. Tech.* **2003**, *140*, 243–247. [CrossRef]
21. Hu, W.; Shin, Y.; King, G. Micromachining of Metals, Alloys, and Ceramics by Picosecond Laser Ablation. *J. Manuf. Sci. Eng.* **2010**, *132*, 011009. [CrossRef]
22. Hocheng, H.; Tai, N.H.; Liu, C.S. Assessment of ultrasonic drilling of C/SiC composite material. *Compos. Part A Appl. Sci. Manuf.* **2000**, *31*, 133–142. [CrossRef]
23. Lee, T.C.; Chan, C.W. Mechanism of the ultrasonic machining of ceramic composites. *J. Mater. Process. Technol.* **1997**, *71*, 195–201. [CrossRef]
24. Li, Z.C.; Jiao, Y.; Deines, T.W.; Pei, Z.J.; Treadwell, C. Rotary ultrasonic machining of ceramic matrix composites: Feasibility study and designed experiments. *Int. J. Mach. Tools Manuf.* **2005**, *45*, 1402–1411. [CrossRef]
25. Bertsche, E.; Ehmann, K.; Malukhin, K. Ultrasonic slot machining of a silicon carbide matrix composite. *Int. J. Adv. Manuf. Technol.* **2013**, *66*, 1119–1134. [CrossRef]
26. Wang, J.; Zhang, J.; Feng, P. Effects of tool vibration on fiber fracture in rotary ultrasonic machining of C/SiC ceramic matrix composites. *Compos. Part B Eng.* **2017**, *129*, 233–242. [CrossRef]
27. Jones, A.H.; Trueman, C.S.; Dobedoe, R.S.; Huddleston, J.; Lewis, M.H. Production and EDM of Si_3N_4–TiB_2 ceramic composites. *Br. Ceram. Trans.* **2001**, *100*, 49–54. [CrossRef]
28. Hu, B.; Yang, Z.; Yang, Z. Research status and development trend of composite machining. *Aerosp. Mater. Technol.* **2000**, 24–31. [CrossRef]
29. Spur, G.; Holl, S.E. Ultrasonic assisted grinding of ceramics. *J. Mater. Process. Technol.* **1996**, *62*, 287–293. [CrossRef]
30. Uhlmann, E.; Spur, G. Surface Formation in Creep Feed Grinding of Advanced Ceramics with and without Ultrasonic Assistance. *CIRP Ann.* **1998**, *47*, 249–252. [CrossRef]
31. Rösiger, A.; Goller, R.; Langhof, N.; Krenkel, W. Influence of in-plane and out-of-plane machining on the surface topography, the removal mechanism and the flexural strength of 2D C/C-SiC composites. *J. Eur. Ceram. Soc.* **2021**, *41*, 3108–3119. [CrossRef]
32. Ding, K.; Fu, Y.; Su, H.; He, T.; Yu, X.; Ding, G. Experimental Study on Ultrasonic Assisted Grinding of C/SiC Composites. *Key Eng. Mater.* **2014**, *620*, 128–133. [CrossRef]
33. Hu, M.; Ming, W.; An, Q.; Chen, M. Experimental study on milling performance of 2D C/SiC composites using polycrystalline diamond tools. *Ceram. Int.* **2019**, *45*, 10581–10588. [CrossRef]
34. Li, C.; Piao, Y.; Meng, B.; Hu, Y.; Li, L.; Zhang, F. Phase transition and plastic deformation mechanisms induced by self-rotating grinding of GaN single crystals. *Int. J. Mach. Tools Manuf.* **2022**, *172*, 103827. [CrossRef]
35. Liu, Q.; Huang, G.; Cui, C.; Tong, Z.; Xu, X. Investigation of grinding mechanism of a 2D Cf/C–SiC composite by single-grain scratching. *Ceram. Int.* **2019**, *45*, 13422–13430. [CrossRef]
36. Liu, Y.; Quan, Y.; Wu, C.; Ye, L.; Zhu, X. Single diamond scribing of SiCf/SiC composite: Force and material removal mechanism study. *Ceram. Int.* **2021**, *47*, 27702–27709. [CrossRef]
37. Ning, F.; Wang, H.; Cong, W. Rotary ultrasonic machining of carbon fiber reinforced plastic composites: A study on fiber material removal mechanism through single-grain scratching. *Int. J. Adv. Manuf. Technol.* **2019**, *103*, 1095–1104. [CrossRef]
38. Xue, F.; Zheng, K.; Liao, W.; Shu, J.; Dong, S. Investigation on fiber fracture mechanism of c/sic composites by rotary ultrasonic milling. *Int. J. Mech. Sci.* **2021**, *191*, 106054. [CrossRef]
39. Tian, L.; Fu, Y.; Xu, J.; Li, H.; Ding, W. The influence of speed on material removal mechanism in high speed grinding with single grit. *Int. J. Mach. Tools Manuf.* **2015**, *89*, 192–201. [CrossRef]
40. Li, C.; Piao, Y.; Zhang, F.; Zhang, Y.; Hu, Y.; Wang, Y. Understand anisotropy dependence of damage evolution and material removal during nanoscratch of MgF_2 single crystals. *Int. J. Extrem. Manuf.* **2022**, *5*, 015101. [CrossRef]
41. Azarhoushang, B.; Tawakoli, T. Development of a novel ultrasonic unit for grinding of ceramic matrix composites. *Int. J. Adv. Manuf. Technol.* **2011**, *57*, 945–955. [CrossRef]
42. Singh, R.; Khamba, J.S. Investigation for ultrasonic machining of titanium and its alloys. *J. Mater. Process. Technol.* **2007**, *183*, 363–367. [CrossRef]
43. Yu, S.J.; Chen, Z.F.; Wang, Y.; Luo, R.Y.; Cui, S. Effect of fabric structure on the permeability and regeneration ability of porous SiCf/SiC composite prepared by CVI. *Ceram. Int.* **2019**, *45*, 11564–11570. [CrossRef]
44. Bansal, N.P.; Lamon, J. *Ceramic Matrix Composites: Materials, Modeling and Technology*; Wiley: Hoboken, NJ, USA, 2014.
45. Zhou, Y.; Tian, C.; Li, H.; Ma, L.-j.; Li, M.; Yin, G.Q. Study on removal mechanism and surface quality of grinding carbon fiber toughened ceramic matrix composite. *J. Braz. Soc. Mech. Sci. Eng.* **2022**, *44*, 476. [CrossRef]

Disclaimer/Publisher's Note: The statements, opinions and data contained in all publications are solely those of the individual author(s) and contributor(s) and not of MDPI and/or the editor(s). MDPI and/or the editor(s) disclaim responsibility for any injury to people or property resulting from any ideas, methods, instructions or products referred to in the content.

MDPI AG
Grosspeteranlage 5
4052 Basel
Switzerland
Tel.: +41 61 683 77 34

Micromachines Editorial Office
E-mail: micromachines@mdpi.com
www.mdpi.com/journal/micromachines

Disclaimer/Publisher's Note: The title and front matter of this reprint are at the discretion of the Guest Editor. The publisher is not responsible for their content or any associated concerns. The statements, opinions and data contained in all individual articles are solely those of the individual Editor and contributors and not of MDPI. MDPI disclaims responsibility for any injury to people or property resulting from any ideas, methods, instructions or products referred to in the content.

www.ingramcontent.com/pod-product-compliance
Lightning Source LLC
Chambersburg PA
CBHW041951260926
51850CB00019B/2469
* 9 7 8 3 7 2 5 8 6 0 8 7 6 *